HUMAN ANATOMY - REVIEWS AND MEDICAL ADVANCES

Edited by **Alina Maria Sisu**

Human Anatomy - Reviews and Medical Advances
http://dx.doi.org/10.5772/66031
Edited by Alina Maria Sisu

Contributors

José Aderval Aragão, Francisco Prado Reis, Felipe Matheus Sant'Anna Aragão, Iapunira Catarina Sant'Anna Aragão, Sydney Correia Leão, Pascal Fallavollita, Zamzuri Idris, Faruque Reza, Jafri Malin Abdullah, Gabriel Cismaru, Andrei Cismaru, Nuket Gocmen Mas, Ayla Kurkcuoglu, David Mazensky, Slavka Flesarova, Gareth Jones

Published by InTech
Janeza Trdine 9, 51000 Rijeka, Croatia

Publishing Process Manager Edi Lipovic
Technical Editor SPi Global
Cover InTech Design team

First published November, 2017
Printed in Croatia
Legal deposit, Croatia: National and University Library in Zagreb

Additional hard copies can be obtained from orders@intechopen.com

Human Anatomy - Reviews and Medical Advances, Edited by Alina Maria Sisu
p. cm.
Print ISBN 978-953-51-3611-8
Online ISBN 978-953-51-3612-5

For EU product safety concerns:
IN TECH d.o.o., Prolaz Marije Krucifikse Kozulić 3, 51000 Rijeka, Croatia,
info@intechopen.com or visit our website at intechopen.com.

Meet the editor

Alina Maria Şişu has been working in the Department of Anatomy and Embryology since 1999 and from 2007 to the English, Medicine, and Dentistry section.
She has written 42 books and chapters from anatomy (anthropology, clinical anatomy), embryology, and pathology field in Romania and abroad. Out of them, three were written in French and six in English.
She has published over 450 articles and abstracts, and participated in 104 meetings, conferences, and congresses.
Her fields of interests are microscopic anatomy, dissection methods, anatomical structures, preservation methods, orthopedics, pathology, embryology, and anthropology.

Contents

Preface

Anatomy is a fundamental science that studies the structure of the human body from ancient times. Over time, the discipline constantly expands with recent progress that has been produced in researching the human body. So, new methods of researching were incorporated in the anatomy development: plastic materials injections, plastination, computed techniques of sectional bodies, and embryology. Anatomic sections like macroscopic, mesoscopic, microscopic, and public anatomies; radiologic anatomy; computed anatomy; radiologic anatomies; and clinical anatomy contribute to realize a very complex discipline that represents the base of learning medicine. "Anatomia clavus et clavis medicinae est." (Anatomy is helm and key of medicine).

The aim of this book is to bring together valuable anatomists/researchers from worldwide in order to sum up modern and advanced techniques of learning and teaching anatomy.

The selected articles highlight new and modern topics concerning anatomy, introducing different approaches of recent entities.

The eight chapters have been selected for their innovative and representative view over the anatomy relevance in the field of medicine, thought as basic and mandatory discipline in medical schools.

The first chapter, "Anatomy in Public: Science, Ethics, and Culture (Review)," introduces the medical teaching instruments, like scalpels, dissection, preservation methods, and the relation between dead body with the once-living individual and their families. To the general public, it is represented by the enormously popular public exhibitions of plastinated cadavers and body parts. The chapter presents the struggle of the anatomists during time, in order to achieve the knowledge of the human body and also the exposed anatomy as plastinated bodies.

The second chapter, "Innovative Technologies for Medical Education,"aims to assess the current practices of anatomy education technology and provides future directions for medical education. It begins by presenting a historical synopsis of the current paradigms for anatomy learning followed by listing their limitations. Then, it focuses on several innovative educational technologies, which have been introduced over the past years to enhance the learning. These include e-learning, mobile apps, and mixed reality. The chapter concludes by highlighting future directions and addressing the barriers to fully integrating the technologies in the medical curriculum. As new technologies continue to arise, this process-oriented understanding and outcome-based expectations of educational technology should be embraced. With this view, educational technology should be valued in terms of how well the technological process informs and facilitates learning and the acquisition and maintenance of clinical expertise.

The third chapter, "Anatomical Correlation of the Arterial Blood Supply to the Spinal Cord in Human and Experimental Animals: A Review,"represents a structured description of animal models existing to examine physiological and functional changes after the spinal cord injury with the aim to explain knowledge about the spinal cord injury in human. The aim of this review is to summarize the available literature into one coherent format. This chapter compares the arterial spinal cord blood supply of the frequently used species (pig, dog, cat, rabbit, and rat) in experimental spinal cord injury and in humans. A complete understanding of the anatomy of the arterial blood supply to the spinal cord is critical for the anatomists and clinicians to determinate the advantages and disadvantages of each animal model for next studies.

The forth chapter, "Human Brain Anatomy: Prospective, Microgravity, Hemispheric Brain Specialization, and Death of a Person," highlights the special features made as a possible site for seat of human soul and form a crucial part in discussion related to death of a person. Besides exploring deep anatomical areas of the brain, superficial cortical areas were also studied. This whole chapter covers superficial, integrative, and deep parts of human brain anatomy with emphasis on brainwaves, brain functions, seat of human soul, and death.

The fifth chapter, "Anatomy of Extramuscular Soleus Veins:Clinical Impact,"provides an anatomical description regarding the venous system of the lower limbs. Their morphologic variability is also detailed in the article and compared with other authors' work. Detailed anatomical knowledge is required for early diagnosis using noninvasive ultrasound techniques. Authors conclude that there is a wide variability in the distribution of soleus veins through the soleus muscle and its quadrants.

The sixth chapter, "Cardiac Anatomy for the Electrophysiologist with Emphasis on the Left Atrium and Pulmonary Veins," aims to provide basic anatomical knowledge for the interventional electrophysiologists, to understand catheter placement and ablation targets. It begins with the location of the heart inside the mediastinum, position of cardiac chambers, pericardial space, and neighboring structures of the heart and continues with the right atrium and important structures inside it: sinus node, cavo-tricuspid isthmus, Koch's triangle, and interatrial septum with fossa ovalis. A special part of this chapter is dedicated to the left atrium and pulmonary veins with the veno-atrial junction, important structures for catheter ablation of atrial fibrillation. Both ventricles with outflow tracts and the coronary venous system are also described.

The seventh chapter, "Anatomical, Biological, and Surgical Features of Human Basal Ganglia,"describes several aspects of the human basal ganglia features. Neurological diseases are characterized through the obvious pathology of the basal ganglia, and there are important findings explaining striatal neurodegeneration on the human brain. Some of these diseases are induced by bacterial and/or viral infections. A comprehensive understanding of the striatal projection loss while receiving striatal input/output on the neurons will contribute to the available knowledge related to pathogenesis of the neurological diseases. Surgical interference can be one alternative for neuronal disease treatment like Parkinson's disease or thiamine-responsive basal ganglia disease or Wilson's disease, respectively, in addition to the vascular or tumor surgery within this area. At the same time, in different pathological processes such as kernicterus, Tourette's syndrome, hemiballismus, obsessive compulsive disorder, neonatal and lacunar infarction, and Huntington's and Parkinson's diseases, were basal ganglia neurons affected.

The eight chapter, "Mesencephalon Anatomy,"deals withthe smallest part of the brain, the midbrain, which serves important functions in motor movement, particularly movements of the eye, and in auditory and visual processing. In addition, its substantia nigra is closely associated with motor system pathways of the basal ganglia. Dopamine is produced in the substantia nigra, and it plays a role in excitation and motivation. The mesencephalic syndromes cause tremor, spastic paresis or paralysis, opisthotonos,nystagmus, and depression or coma. In addition, cranial trauma, brain tumors, thiamine deficiency, and inflammatory or degenerative disorders of the mesencephalon have also been associated with the midbrain syndrome.

The book is the result of many researchers' sustained work. I gratefully acknowledge the assistance provided by all the authors that have contributed to this book publication and by the InTech editorial office that initiated this project completed by the book edition.

Alina Maria Şişu, MD, PhD
Senior Lecturer
Department of Anatomy and Embryology,
"Victor Babeş" University of Medicine and Pharmacy,
Timişoara, Romania

Section 1

Reviews and Medical Education

Chapter 1

Human Anatomy: A Review of the Science, Ethics and Culture of a Discipline in Transition

David Gareth Jones

Additional information is available at the end of the chapter

http://dx.doi.org/10.5772/intechopen.68524

Abstract

Anatomy has undergone radical changes over its history, and even now its appearance varies between audiences. Within academia, it has frequently been seen as the bastion of medical teaching, even as a handmaid of surgery. To the general public over recent years, it is represented by the enormously popular public exhibitions of plastinated cadavers and body parts. Increasingly within medical teaching, it has acquired a far more humanistic face, epitomized by ceremonies at the start and end of dissection to connect the dead body with the once living individual and his/her families. Modern anatomy has also developed a strong research ethos. These movements can be traced in the many editions of *Gray's Anatomy*, from 1858 to the present day. However, the humanistic side of anatomy reminds us that anatomy is not merely a science, since its ethical dimensions are legion as it has transformed from a dubiously moral and barely legal activity to one that now aims to manifest the highest of ethical standards. Nevertheless, it continues to have challenging dimensions, such as its ongoing dependence upon the use of unclaimed bodies in many societies. These challenges are reminders that anatomy does not remain stationary.

Keywords: *Gray's Anatomy*, culture of dissection, humanistic face of anatomy, commemoration ceremonies, plastination, *Body Worlds*, ethical guidelines

1. Introduction

For some, the discipline of anatomy is characterized by formalin cadavers in sterile dissecting rooms and very large amounts of detailed anatomy. Within academia, it has frequently been seen as the mainstay of medical teaching, even as a handmaid of surgery. To others, it has a far more humanistic face, as demonstrated by ceremonies at the start and end of dissection to connect the dead body with the once living individual and their families. To the general public, it is represented by the enormously popular public exhibitions of plastinated cadavers and body parts. However, these descriptions amount too little more than facets of what constitutes anatomy in modern guise, with its strong research ethos and broad scope from biological anthropology and clinical anatomy to molecular biology and genetics. The one unitary theme across this broad swathe of biomedical endeavor is structure or organization, the fundamental thread that ties together all these approaches within modern anatomical science. These movements can be traced through the many editions of *Gray's Anatomy*, from 1858 to the present day.

However, the humanistic side of anatomy reminds us that anatomy is more than merely a science. Its ethical dimensions are as numerous as its scientific credentials as it has transformed from a questionably moral and legal activity to one that now aims to manifest the highest of ethical standards, even though in many societies it continues to have challenging dimensions, such as its ongoing dependence upon the use of unclaimed bodies. In these and other ways, anatomy has entered uncharted territory with previously unexplored ethical dimensions.

2. Setting the scene: *Gray's Anatomy*

The anatomists' core text, *Gray's Anatomy*, reflects the many dimensions of anatomy. The 41st edition of the English version, published in 2016, 156 years after the first edition, is both impressive and near exhaustive in its coverage. Its major sections range from Cells, Tissues and Systems, and Embryogenesis, to the regional subdivisions of the human body [1]. The visual impression made on the reader is of high-quality illustrations, with their dependence upon a range of contemporary techniques, from classic histology to immunofluorescence and immunolabelling. This is the traditional anatomists' approach in contemporary form. But much of the detail is worlds removed from what was available 50 years ago, let alone 150 years ago. The moniker, 'The Anatomical Basis of Clinical Practice' appeared first in 2005 with the 39th edition, emphasizing that anatomy is to be viewed within a health sciences context, since this is a major driver for understanding and appreciating the anatomical organization of the human body. It is also important to note that from the 1973 edition onwards, there has been a willingness to admit that there are gaps in our knowledge of anatomy, including gross anatomy, and that our understanding of the human body is far from all-knowing [2]. In other words, ongoing research is vital if anatomy at all levels is to keep abreast of developments in allied biomedical disciplines, a thrust that subsequent editions of *Gray's Anatomy* have attempted to continue.

While it is easy to be captivated by a modern edition of this classic text, its beginnings lay in the 1850s in mid-Victorian London. That was an era of immense cultural ferment in both literary and scientific fields [2]. Surgery, too, was making great strides now that anesthesia was becoming increasingly available. It was a time of excitement and ferment, when two young medical men, Henry Gray and Henry Vandyke Carter, who was also an accomplished artist, were getting together on the project that was to result in the publication of *Anatomy Descriptive and Surgical* [3]. Little is known about Gray himself, except that he was an up-and-coming surgeon, who died from smallpox 3 years after the publication of his *Anatomy*, at the age of 34.

Two points are worthy of note. The first is that Gray was a researcher [4]. He had published papers on the histology, embryology and comparative anatomy of the anatomical origins of the optic nerves, and later on the spleen. This latter work was published as his first book: *The Structure and Functions of the Human Spleen*. It is clear that research was an important foundation for the work he did a short time later on his magisterial text.

Second, Gray and Carter themselves carried out the dissections of bodies at St. George's Hospital in London. The bodies would have been those of the poor from workhouses, prisons, and hospitals, whose remains had not been claimed by relatives. They were unclaimed, although since misconduct was rife and few formal records were kept, in all likelihood it was deception that led to some ending up as unclaimed. Nowhere in *Gray's Anatomy* is their origin or predicament mentioned. While this is typical rather than atypical of anatomy texts, both in the mid-nineteenth century and much later, it points to a gap between the stunning illustrations of normal human anatomy and the sources of the bodies that provided the raw material for the illustrations. Historian Ruth Richardson [4] has commented: "In *Gray's*, the legally sanctioned bodies of people utterly alone in the metropolis were the raw material for dissections that served as the basis for illustrations that were rendered in print as wood gravings. As mass-produced images, they have entered the brains of generations of the living—via the eyes, the minds, and the thoughts of those who have gazed at them." (p. 139) There is no memorial to those whose bodies provided so much for generation after generation of anatomists and students of anatomy.

There is no evidence of wrong doing on the part of these two men. Nevertheless, they serve as a reminder that the culture in which modern clinical anatomy was born was far removed from the expectations of the twenty-first century. By the same token, they were also far removed from the culture in which anatomy as a modern enterprise was born three centuries earlier during the European Renaissance [5]. Anatomy is, therefore, an evolving discipline, much like any other. Anatomy today should look different from 50 years ago, and it will be different again in 50 years' time. Anatomy also assumes different forms in different countries.

3. Anatomy and the culture of dissection

Anatomy as a science emerged during the Renaissance, as it strove to attain its own niche within the spectrum of emerging academic disciplines. If the verb 'anatomize' and the noun 'anatomization' were to be employed today, they would be used in a scientific sense. This is because the anatomization or dissection of a body reduces it to its component parts in an effort to construct a new body of knowledge. In light of that which is learned about specific bodies, the intention is to strengthen and broaden the science of anatomy in general [6].

Although the term 'body,' in its primary usage, refers to the body of human beings or of animals, it is an abstraction. We move continually between a particular body, *somebody's* body, and the body in general [7]. Cadavers, body parts, tissues and bony remains always come from particular individuals, and even when these individuals lived in the distant past, they can never be completely dehumanized.

In sixteenth- and seventeenth-century Europe, the culture of dissection emerged out of a bewildering array of competing cultural forces. Prior to the modern, dispassionate, scientific approach to the human body, anatomy was part of a popular culture fascinated by the interior of the body but unable to delve into this largely uncharted domain [5, 6].

There was widespread fascination with the body, and this led to morbid curiosity with dissection, since this was the only way of exploring the interior of the body. However, since the bodies had come directly from the gallows, there was a close association between the anatomist and the executioner. The end result was that the criminal, the executioner, and the anatomist, each had a role to play in what has been termed 'the culture of dissection' [5]. The three were interlocked in this macabre process, in which there was neither the notion nor hint of clinical detachment.

Anatomists, therefore, found themselves active participants in the execution process, even appearing to be accomplices of the executioners. This was essential if the supply of human bodies was not to dry up. But this was a problem, since it sent a clear message to the general public that anatomists were much closer to criminals than to the respectable members of society. To overcome this, their activities were bestowed with divine significance, and so as they investigated the usually unapproachable realm of the interior of the body, they were looking into what was in effect a sacred temple [5]. In this manner, the status of anatomists was placed on a par with divine activities thereby elevating them above the level of criminals.

A related phenomenon was the category of self-dissection, epitomized by Andreas Vesalius among others [8]. As the name suggests, the dead body was depicted as being actively involved in the dissection process itself. This hinted that the body may not be as dead as one would suppose, since anatomy was animating the body and endowing it with a life of its own. The end result was the impression that knowledge of the (dead) body was actually knowledge of the living, thereby stressing the naturalness of dissection [5].

In spite of these subtle moves, the dissection of cadavers remained problematic. It still existed on the edge of living society, with dissected cadavers being seen as a disturbing community of the dead. In depicting cadavers like this, Vesalius and others were claiming that the anatomist was not disrupting the body, but was assisting the natural process of decay [5]. In their different ways, all were attempts to eliminate the gulf between the dead and the living, moves that have reappeared in recent times in the public displays of plastinated bodies (see *Anatomy exposed to public gaze—plastination*). This suggests that societies' unease at dissection and the use of the dead body continues to manifest itself, no matter how much the circumstances surrounding it have changed.

Creative as were these attempts to overcome concerns about the work of anatomists during the Renaissance period, doubts remained. The whole process of dissection was accompanied by horror and fascination, especially on the part of writers and poets. The result was the morbid eroticism of some Renaissance poetry and theatre, in which writers sought to explore the unknown mysteries of the body's interior, with erotic dreams of dissection (such as the 1659 poems of Richard Lovelace; see Ref. [9]). Strange as these works appear to contemporary anatomists, they point to an abiding truth, namely, that anatomy is never carried out in a cultural and philosophical vacuum, regardless of the culture implicated (See Ref. [5] for examples from the Renaissance period in Europe, especially Chapters 3, 4 and 5).

The transition from these fascinating but perplexing times to the late eighteenth and early nineteenth centuries was fraught, as the demand for bodies for dissection continued to outstrip the legal means of supply. The stage was set for the unsavory next stage in the questionable beginnings of modern anatomy: use of the bodies of executed criminals, body snatching and occasionally murder [10]. The world of the so-called resurrectionists and the host of macabre stories about the indecent haste with which the recently buried were transported from graveyard and poorhouse to anatomy dissecting room are ethically foreign and deeply embarrassing to the world of contemporary anatomy [11, 12].

The pivotal 1832 Anatomy Act in England proved ground breaking by introducing into the anatomical lexicon the concept of unclaimed bodies. It was both revolutionary and disconcerting, since it made available large numbers of otherwise inaccessible bodies, but in doing so ensured that most of these would be those of the poor [10]. The lack of any incentive to revisit this decision, by not considering the alternative of soliciting bequests, ensured that for many years into the future, there was widespread willingness among anatomists to make use of the bodies of the disadvantaged and dispossessed. This lack of ethical reflection legitimized the unclaimed paradigm as the normal source of bodies for anatomical investigation. This, in turn, opened the doors to widespread use of the bodies of the mentally ill, of African-Americans, and of those executed in concentration camps during the Nazi era [13–15].

The result for anatomists has been tension between the legitimate scientific desire to work on high-quality material (fresh material obtained shortly after death), and the ethical imperative of soliciting informed consent from a donor prior to death. While all uses of unclaimed bodies do not fit into the outrageous categories referred to above, and while there is ongoing debate about the precise nature of informed consent [16, 17], lack of any reference to the centrality of informed consent has cast a pall over the ethical environment of anatomy as a discipline. This will only be rectified as anatomy explicitly argues for, and implements, the ethical superiority of body bequests over unclaimed bodies [18].

These historical allusions all attest to the assertion that anatomy, and especially gross anatomy, is not a self-justifying regime. It is not carried out in a cultural and philosophical vacuum, and this affects every aspect of human anatomy. Research on human embryos may be regarded as the face of this debate today. Subjecting human embryos to research procedures requires the assent of the communities in which these are being conducted [19]. This work is no more self-justifying than is the use of human cadavers as a source of organs.

4. The humanistic face of anatomy

The discussion so far has been concerned with the way in which anatomy has gained a foothold in the scientific arena, giving it a legitimate stake in investigations on the structure of the human body. This underlies its potential contributions in both research and teaching terms. Avenues open to anatomists in research investigations are entirely dependent upon its scientific credentials, and until more recent times, this has also been the case in teaching the fundamentals of anatomy, at both the gross and microscopic levels. However, it has become evident that to confine anatomy to its scientific dimensions in teaching health science students have limitations, since these students will be entering professions in which empathy with patients is paramount.

As this realization has increasingly taken root in the thinking of educators, there has been a major move in the direction of seeing anatomy as much a humanistic discipline as a scientific one [20]. This is not a rejection of the scientific basis of anatomy, but an attempt to place it within a patient-centered health science context. One manifestation of this is in the emergence of commemorations and memorials in association with the donation of bodies for medical education [21–24].

Holding ceremonies to acknowledge and thank the families of those who have donated their bodies to anatomical education is recognition that anatomists are an integral part of their communities and are dependent upon the goodwill of others [25]. This sends the message that anatomists are human beings dealing with the remains of fellow human beings, and that anatomy as a discipline takes account of this relationship. These ceremonies explicitly acknowledge that the bodies available for study have been donated for this purpose by people who gave their fully informed consent. This, in turn, emphasizes the centrality of body bequests in ethical thinking within anatomy [18, 26].

The variety of terms used to convey the essence of these ceremonies include commemoration, thanksgiving, ceremony, service, and memorial ('memorial ceremonies' [23, 27]; 'Convocation of Thanks' [22]; 'Thanksgiving Service' [28]; 'cremation/burial ceremony' [29]). In their different ways, each conveys the notion of remembrance, and of paying tribute to those who in their death have donated their bodies to a worthy cause, that of medical teaching and research [25].

The donors are remembered for what they have given, and a ceremony is one public manifestation of this gift of inestimable value. Their altruism is recognized and saluted, and their families are thanked for the support they have provided in enabling the giving of this gift. Giving something closer to oneself than anything else also signifies trust in the anatomy staff and students, in the expectation that their bodies will be treated with respect and dealt with it in a manner worthy of the donor's memory.

Ceremonies point toward the humanistic face of anatomy, and the unacceptability of treating cadavers merely as research and teaching tools. Their social and cultural context frames all facets of anatomical study and of the display of human remains. If this is now recognized as a central feature of anatomy, it becomes important to ask where the large public displays of plastinated dissected human bodies fit. These, after all, have become an indelible face of anatomy but what message do they convey about the character of anatomy? Do they have a humanistic face as I have been arguing, or is their rationale purely scientific *a la Gray's Anatomy*?

5. Anatomy exposed to public gaze: plastination

These vast plastinated exhibitions and their place within the world of anatomy have been described and assessed by numerous commentators (see Ref. [30] for references). They are directed at the general public and not at medical and other heath science academics and students. They are intended to take anatomy out of the secretive dissecting room and into the public arena. This is what has been referred to by Gunther von Hagens, the founder of the *Body Worlds* empire, as the 'democritization' of anatomy [6, 31], the release of anatomy from its privileged position within the halls of academia and into the wider world. Knowledge of one's own body is seen as something that everyone should have access to, and displaying the body in its dissected state is the ideal way of accomplishing this. It is only in this way that people can begin to appreciate what organs look like, how they relate to each other, along with the vessels and nerves that supply them. But how is this to be done, since a replica of the dissecting room with its smell of formalin-impregnated death and lifeless preserved cadavers on slabs would hardly attract a wide audience?

The breakthrough came when von Hagens devised a new method of preservation of human tissues, plastination, in which tissue fluids are replaced with plastic [32, 33]. This was a major step forward for use in teaching human anatomy to health science students, where it is used to preserve previously dissected body parts. Additionally, it proved beneficial for research purposes with the use of body slices. These uses of course are confined to academia.

The move to public displays came with the preservation, not of body parts, but of whole bodies that are referred to as 'plastinates.' But more significantly, rather than being displayed horizontally, they are shown vertically. Not only this, they can be depicted in a variety of stances, and to give the impression of running, walking and jumping, playing a variety of sports, and even having sexual intercourse [34]. Using these devices they appear to be 'alive,' far removed from the lifeless inactivity of the dissecting room. They may be dissected, but they give the impression of participating in the vigorous life of everyday existence. The effect is frequently dramatic and awe inspiring, and prompts reactions of wonderment at the beauty and complexity of the human body. For some, this positive side is matched by complete rejection on the ground that they are a travesty: disgusting, disconcerting, demeaning, and dehumanizing [30]. What has been fascinating is that initial objections to them have come mainly from anatomists [35, 36] and religious leaders [37–39]. While more in-depth analyses have dispelled some of this negativity [40–42], the impassioned responses of some individuals have uncovered wellsprings of unease [43].

For others, it is their attractiveness and aesthetic beauty that have proved a drawcard for millions of people worldwide. Of these, *Body Worlds* is the best known on account of its leadership within the field, the high profile of its founder, Gunther von Hagens, its leading-edge technology and the very high quality of its dissections. Moreover, von Hagens' philosophical claims regarding the status of plastinates have been the driving force for much academic comment [43–46]. In line with this, the exhibitions have occasioned a considerable body of scholarly work from many different disciplines, touching on the haunting ways in which the bodies are displayed, impressions left by the exhibitions, and the legitimacy of investigations on the human body as an object of scientific curiosity [47–50].

The question that looms large over plastinates is their nature. What are they? At one level, they are simply dead human beings, dissected in a variety of interesting ways. But this is a superficial response, since the method of preservation has fundamentally altered their tissue that now makes up no more than 30% of the body. The remaining 70% is plastic, raising the question of how this hybrid entity relates to the 'normal' human body. Ambivalence has crept in. The plastinate is more than a plastic model of a human body, as it still contains human tissue that mirrors important facets of that particular individual during life. No two plastinates are identical, any more than two individuals are identical. They have been modified to create a new entity, based on a human template but increasingly artificial [43]. The end result is an enigma, because while plastinates are allegedly about the dynamic and living body, the newly constructed plastinated body is far removed from that of the original living individual. They represent their own category of what may be described as 'living deadness,' occupying a 'post-mortal world,' part mortuary and part art gallery [43, p. 191]. They are dead, and yet the process of plastination ensures that they will not decay; they are frozen in some intermediate state [51].

These quandaries are made far more troublesome by the way in which they are exhibited for the general public, as though they are experiencing some ongoing existence, a form of post-mortal existence [42, 52–54]. How can you play basketball if you are dead, and yet some of these plastinates are depicted as doing just this? This can be dismissed as poetic license; this is merely an exhibition, and it is unlikely that any of those viewing them will think they are actually alive. That is true, and yet the apparent 'immortality' of plastinates has been plugged as an important aspect of the whole venture [52, 55], an emphasis that has proved immensely problematic for anatomists and others [56]. Even though 'immortality' is an exaggeration, it seems to represent a new category of human body, separate from both fresh corpse and decaying remains [6].

King et al. [30] argue that plastinates do not occupy standard cultural binary categories such as interior or exterior, real or fake, dead or alive, bodies or persons, and self or other. This is because they transgress the usual boundaries by which we describe and understand the world. They refuse to be pigeon-holed, no matter how hard we try. Even the simplest designation of dead or alive escapes us [30, 54]. They are representations of real bodies [46, 57], having been modified to produce something that is an artificial representation of perfected nature [48]. The artistic component is essential to the success of the end-product, but this removes it from the sphere of vulnerability and imperfection that characterizes human existence, a vulnerability resulting from biological, environmental and social factors as well as from moral and spiritual ones [58].

This lack of clarity regarding their categorization surfaces repeatedly and has enabled Von Hagens to employ the description 'post-mortals' [6, 53]. The lack of identification with the person who once lived, with no trace of their values, attitudes or ideas, reduces them to depersonalized bodies. Even the traces of memory by which someone lives on have been defaced, since in the absence of discernible external bodily features, there is no way in which relatives and friends can recognize the plastinated remains [58]. For one commentator, the absence of a personality, friends, family and history leaves a gaping and eerie vacuum that forcefully calls into question what it is to be human and reminds us of what few of us like to dwell on—our mortality. They are 'bodies with no soul' [59].

What emerges very clearly from this discussion is the impression created by the exhibitions, namely that anatomy is science, no more and no less, with the bodies on display representing the generalized and abstracted body. It has nothing to do with the humanistic trends increasingly being manifested in contemporary anatomy. Neither the basketball and football players, nor the ballerinas, in the exhibitions are real people; they are representatives of these sports and activities. We do not know whether the individuals represented by these bodies ever indulged in these activities. They are being portrayed to tell a story that, in all probability, has nothing to do with the people prior to their death and plastination and subsequent display in these exhibitions. In other words, they tell us nothing about real people with real life histories. This is far removed from the humanizing trajectories that seek to enhance students' relationships with the bodies they are dissecting [60]. Students cannot develop a relationship with a plastinate, no matter how useful it is in helping them follow nerves and blood vessels [61].

Far from humanizing the body, these public exhibitions appear to distance themselves from the people who have been plastinated, and in doing this, they objectify the body. They dislocate the body from a clinical and relational base, since they have removed them from the environment that nurtured them and of which they were an integral part. Plastinates do not represent the bodies of somebody, but have been generalized to represent bodies in general. This is acceptable as anatomy *per se*, the traditional anatomy of *Gray's Anatomy*, presenting the data that medical students and others have to learn, but it fails to engender any humanistic element in the anatomy. It is misleading therefore to label any of the exhibitions as depicting 'real' people; they are real dissected bodies, useful in some ways but only part of the story of anatomy.

6. Anatomy and its ethical dimensions

The developments in anatomical thinking over many years, and especially over recent ones, have led to concerted efforts to raise the profile of ethical thinking as a basis for anatomical thinking and investigations. While this has been undertaken by individual anatomists, it has also been taken on board by anatomical societies representing anatomists from across the globe. These societies are represented by the International Federation of Associations of Anatomists (IFAA) that has formulated a set of ethical guidelines with a view to overseeing the donation of human bodies [62, 63].

Procedures of the highest ethical standards are required, in order to give donors full confidence in their decision to donate. This in turn demands trust on the part of the public. The guidelines are as follows. The underlying premise is that bodies have been bequeathed for teaching and research purposes.

1. Informed consent from donors must be obtained in writing before any bequest can be accepted. Consent forms should take into account the following:

   a. Donors must be entirely free in their decision to donate, this excludes donation by minors and prisoners condemned to death.

   b. Although not essential, good practice is encouraged by having the next of kin also sign the form.

   c. Whether the donor consents to their medical records being accessed.

2. There should be no commercialization in relation to bequests of human remains for anatomical education and research. This applies to the bequest process itself, where the decision to donate should be free from financial considerations, and also to the uses to which the remains are put following bequest. If bodies, body parts, or plastinated specimens are to be supplied to other institutions for educational or research purposes, this may not yield commercial gain. However, charging for real costs incurred, including the cost of maintaining a body donation program and preparation and transport costs, is considered appropriate. Payment for human material per se is not acceptable.

3. There needs to be an urgent move toward the establishment of guidelines regulating the transport of human bodies, or body parts, within and between countries.

4. Specimens must be treated with respect at all times. This includes, but is not limited to, storing and displaying human and non-human animal parts separately.

5. The normal practice is to retain donor anonymity. Any exceptions to this should be formally agreed to beforehand by the bequestee and, if appropriate, the family.

6. Limits need to be placed on the extent to which images, or other artifacts produced from donations are placed in the public domain, including in social media, both to respect the privacy of the donor (and their surviving relatives) and to prevent arousing morbid curiosity. No individual should be identifiable in images.

7. A clear and rigorous legal framework should be established on a national and/or state level. This legal framework should detail:

   a. The procedures to be followed in accepting bequests of human remains for anatomical examination, including who is responsible for human remains after death.

   b. The formal recognition of institutions which may accept bequests, which in some jurisdictions may involve licensing.

   c. The safe and secure storage of human remains within institutions.

   d. The length of time such remains will be retained by the institution.

   e. The procedures to be followed in disposing of remains once the anatomical examination is complete and they are no longer required for anatomical education and research.

8. Institutional procedures should be formally established by an oversight committee, which shall review the body donation program at regular intervals. Such procedures should include the following:

    a. Copies of the bequest should be retained both by the donor and by the institution for whom the bequest is intended.

    b. Records should be kept for a minimum of 20 years from the date of disposal to ensure that human material can be identified as originating from a specific donor.

    c. Good conservation procedures should be employed throughout the entire period during which the human remains are retained to ensure that the most effective use is made of any bequest received.

    d. Efficient tracking procedures should ensure that the identity and location of all body parts from an individual donor are known at all times.

    e. Facilities where cadavers are used must be appropriate for the storage of human remains and secured from entry by unauthorized personnel.

9. There needs to be transparency between the institution and potential donors and their relatives at every stage, from the receipt of an initial enquiry to the final disposal of the remains. The clear communication of information should include but not necessarily be limited to the production of an information leaflet (hard copy and/or digital), which could also help publicize anatomical bequests and increase the supply of donors. This should set out the following:

    a. The procedures relating to registering bequests, acceptance criteria, the procedures to be followed after death (including under what circumstances a bequest might be declined), and the procedures relating to disposal of the human remains. Sufficient grounds for rejection could include, but need not be limited to:

        - the physical condition of the body.
        - the virological or microbiological status of the donor in life.
        - the existence of other diseases (e.g., neurological pathology) that might expose staff or students handling the body to unacceptable risks.
        - body weight or height over a specified limit.
        - the possible over-supply of donations at that institution at that time.
        - place of death outside the designated area from which bodies are obtained.

    b. The range of uses of donated bodies at that institution.

    c. Possible costs, if any, that might be incurred by the bequestee's family in making a bequest, and the costs to be met by the institution accepting the bequest.

    d. Whether the donor's anonymity will be preserved and whether their medical history accessed.

e. Whether the body or body parts might be supplied to another institution.

f. The maximum length of time the body will be retained, including any legally sanctioned possibility of indefinite retention of body parts. The relatives of the donor should be given the option of being informed in due course of the date when the remains will be disposed of.

g. Donors should be strongly encouraged to discuss their intentions with their relatives to ensure that their relatives are familiar with their wishes and that as far as possible those wishes will be carried out after death.

10. Special lectures/tutorials in ethics relating to the bequest of human remains should be made available to all students studying anatomy. This is to encourage the development of appropriate sensitivities in relation to the conduct and respect that is expected of those handling human remains used for purposes of anatomical education and research.

11. Institutions should be encouraged to hold Services of Thanksgiving or Commemoration for those who have donated their bodies for medical education and research, to which can be invited relatives of the deceased, along with staff and students.

The guidelines should not be regarded as having been set in concrete once and for all and are to be modified in light of ongoing ethical reflection. For instance, the anonymity of cadavers has been raised as a matter for discussion [64], while commercialization requires further nuancing in societies where there are for-profit groups alongside not-for-profit ones [65]. Similarly, the transport of human tissue across countries' borders remains a grey area.

Human Tissue Acts that govern the practice of anatomists when dealing with human material have been re-written over recent years in response to a series of organ and body parts scandals. These expectations are now set out in one Act after another and regard informed consent by appropriate parties as a crucial ethical driver (HTAs).

One might have hoped that scandals involving the misuse of dead human bodies would have been consigned to history. However, this has not been the case as epitomized in the most extreme fashion in Germany and allied territories during the Nazi regime [14] and in less extreme ways by organ donor scandals in pathology departments in a number of countries from the 1960s onwards and brought to light around the year 2000 [66–69].

In summarizing the findings to emerge from her magisterial study of anatomy during the Third Reich, Hildebrandt [14] referred to research on the 'future dead,' as one ethical value after another was dispensed with and the profession was converted into 'an agent of evil through the convergence of their own reductionist view of human life, the National Socialist exclusionary medical ethics, and the new "opportunities" provided by the regime' (p. 307). This trajectory involved what she describes as the 'destruction of memory' and the complete annihilation of any professional ethics. For Hildebrandt [14], the take-away message is that 'the benefit for the individual must remain at the center of medical ethics, not the potential benefit for the society as a whole. In that respect, the medical practitioner will always have to take a political stance' (p. 325).

Scandals, let alone rampant examples of evil tarnish the reputation of all who are dependent upon the goodwill and support of the public to obtain the material they require for both teaching and research. Hence, ethical practice assists the profession as a whole by cementing its standing in the eyes of society through recognizing the humanity and individuality of deceased individuals [70].

7. Future dimensions of anatomy

The pivotal contributions of Vesalius and Gray among many others need to be constantly recognized for their seminal contribution to what anatomy is today. And yet they were children of their times, who worked in vastly different environments from each other and from the ones encountered in the twenty-first century. We cannot understand either them or their contributions if we ignore their respective contexts. In the same way, we ourselves cannot be understood apart from our contexts, and we have the ability to change them in at least some respects.

Anatomy does not remain stationary, and neither can the expectations of any one society remain isolated from those of similar and very dissimilar societies. Further, the lessons of history may prove far more relevant to current challenges than could ever have been foreseen. For instance, today, it has to contend with the pressures and opportunities opened up by cyberspace.

One of my dominant concerns is the way in which the availability of anatomical dissections on media, such as YouTube, may normalize public perceptions of anatomy in ways over which the anatomical profession has no control [71]. Whether this will have an effect on the trust that is integral to the relationship between institutions, donors, families, and communities, and crucial for the existence of healthy donor programs, has to be seen. Technology is having profound implications for anatomy as it is for every other health science discipline. Among these are ethical implications, and if its practitioners at large fail to grasp this, the consequences could be deleterious to human welfare.

Author details

David Gareth Jones

Address all correspondence to: gareth.jones@otago.ac.nz

Department of Anatomy, University of Otago, Dunedin, New Zealand

References

[1] Standring S, editor. Gray's Anatomy: The Anatomical Basis of Clinical Practice. 41st ed. London: Elsevier; 2016. p. 1151

[2] Richardson R. Historical introduction. In: Standring S, editor. Gray's Anatomy: The Anatomical Basis of Clinical Practice. 40th ed. London: Churchill Livingstone, Elsevier; 2008. pp. xvii–xxi

[3] Gray H, with drawings by Carter HV. Gray's Anatomy: Anatomy, Descriptive and Surgical. 1st ed. London: JW Parker and Son; 1858

[4] Richardson R. The Making of Mr. Gray's Anatomy. Oxford: Oxford University Press; 2008. p. 322

[5] Sawday J. The Body Emblazoned. London: Routledge; 1995. p. 327

[6] Jones DG, Whitaker MI. Engaging with plastination and the Body Worlds phenomenon: A cultural and intellectual challenge for anatomists. Clinical Anatomy. 2009;**22**:770-776

[7] Kass LR. Thinking about the body. Hastings Center Report. 1985;**15**:20-30

[8] Vesalius A. De Humani Corporis Fabrica. 1543 www.vesaliusfabrica.com/en/original-fabrica/the-art-of-the-fabrica/newly-digitized-1543-edition.html

[9] Wilkinson CH, editor. The Poems of Richard Lovelace, Clarendon Press: Oxford; 1930. p. 132

[10] Richardson R. Death, Dissection and the Destitute. 2nd ed. London: Phoenix Press; 2001. p. 426

[11] MacDonald H. Human Remains: Episodes in Human Dissection. Melbourne: Melbourne University Press; 2005. p. 220

[12] MacDonald H. Possessing the Dead: The Artful Science of Anatomy. Melbourne: Melbourne University Press; 2010. p. 289

[13] Hildebrandt S. Anatomy in the Third Reich: An outline. Part 2. Bodies for anatomy and related medical disciplines. Clinical Anatomy. 2009;**22**:894-905

[14] Hildebrandt S. The Anatomy of Murder: Ethical Transgressions and Anatomical Science during the Third Reich. New York: Berghahn; 2016. p. 374

[15] Jones DG. The anatomy museum and mental illness: The centrality of informed consent. In: Coleborne C, MacKinnon D, editors. Exhibiting Madness in Museums: Remembering Psychiatry through Collection and Display. New York: Routledge; 2011. pp. 161-177

[16] Wilkinson TM. Respect for the dead and the ethics of anatomy. Clinical Anatomy 2014;**27**:286-290

[17] Jones DG. Using and respecting the dead human body: An anatomist's perspective. Clinical Anatomy 2014;**27**:839-843

[18] Jones DG, Whitaker MI. Anatomy's use of unclaimed bodies: Reasons against continued dependence on an ethically dubious practice. Clinical Anatomy. 2012;**25**:246-254

[19] Jones DG. Where does New Zealand stand on permitting research on human embryos? The New Zealand Medical Journal. 2014;**127**(1399). http://journal.nzma.org.nz/journal/127-1399/6236/

[20] Stekalj G. Humanistic Anatomy: A New Program for an Old Discipline. New York: Nova Science Publishers; 2016. p. 69

[21] Hildebrandt S. Developing empathy and clinical detachment during the dissection course on gross anatomy. Anatomical Sciences Education. 2010;**3**:216

[22] Pawlina W, Hammer RR, Strauss JD, Heath SG, Zhao KD, Sahota S, Regnier TD, Freshwater DR, Feeley MA. The hand that gives the rose. Mayo Clinic Proceedings. 2011;**86**:139-144

[23] Jones TW, Lachman N, Pawlina W. Honoring our donors: A survey of memorial ceremonies in United States Anatomy programs. Anatomical Sciences Education. 2014;**7**:219-223

[24] Bolt S. Dead bodies matter: Gift giving and the unveiling of body donor monuments in the Netherlands. Medical Anthropology Quarterly. 2012;**26**:613-634

[25] Jones DG. Bioethical aspects of commemorations and memorials. In: Strkalj G, Pather N, editors. Exploring the Human Face of Anatomy: Commemorations and Memorials in Anatomy Programs. World Scientific Publishing. 2017, in press

[26] Jones DG. Use of bequeathed and unclaimed bodies in the dissecting room. Clinical Anatomy. 1994;**7**:102-107

[27] Zhang L, Wang Y, Xiao M, Han Q, Ding J. An ethical solution to the challenges in teaching anatomy with dissection in the Chinese culture. Anatomical Sciences Education. 2008;**1**:56-59

[28] University of Queensland. UQ body donor Thanksgiving Ceremony [Internet]. 2015. Available from: https://www.uq.edu.au/sbms/thanksgiving-ceremony [Accessed: 2015-09-02]

[29] Lin SC, Hsu J, Fan VY. "Silent virtuous teachers": Anatomical dissection in Taiwan. British Medical Journal. 2009;**339**:b5001

[30] King MR, Whitaker MI, Jones DG. I see dead people: Insights from the humanities into the nature of plastinated cadavers. Journal of Medical Humanities. 2014;**35**:361-376

[31] Institute for Plastination. Donating Your Body for Plastination [Internet]. 2007. Available from: http://www.bodymobil.de/Downloads/Englisch/BD_brochure_ENG_low 120208.pdf [Accessed 2008-10-01]

[32] von Hagens G. Impregnation of soft biological specimens with thermosetting resins and elastomeres. The Anatomical Record. 1979;**194**:247-256

[33] von Hagens G, Tiedemann K, Krit W. The current potential of plastination. Anatomy and Embryology. 1987;**175**:411-421

[34] Body Worlds images [Internet]. 2016. Available from: https://www.google.co.nz/search?q=body+worlds+imnages&ie=utf-8&oe=utf-8&client=firefox-b&gfe_rd=cr&ei=pfS8V4T-Fc_r8AemoYKoCA [Accessed 2016-08-24]

[35] Boyde A, Fraher P, Morris JF, Moxham BJ, Bennett JP, Newell R, Soames RW, Dangerfield P. Letter: Dissections in display [Internet]. The Independent, 16 March 2002. Available from: http://findarticles.com/ p/articles/mi_qn4158/is_20020316/ai_n12600707. [Accessed: 2009-01-21]

[36] Kuhnel W. Statement by the Anatomische Gesellschaft on the Infamous Body-World Show of Dr Gunther von Hagens. Plexus: Newsletter of the International Federation of Associations of Anatomists, December 2004. Available from: http://www.ifaa.net/PLEXUS_DECEMBER_2004-FINAL.pdf [Accessed: 2009-01-21]

[37] Dolce L. Injustice perpetrated on the dead. The National Catholic Bioethics Quarterly. 2010;**10**(4):667-676

[38] Coulson C. Cadaver art: A chamber of horrors, The Colson Center Library. 1998. Available from: http://www.colsoncenter.org/bpcommentaries/breakpoint-commentaries.../13/12904 [Accessed: 2016-05-16]

[39] Metaxas E. Ghoulish 'Art': The Body Worlds Exhibit, The Christian Post. 2012. Available from: http://www.christianpost.com/news/ghoulish-art-the-body-worlds-exhibit-75158/ [Accessed: 2016-05-16]

[40] Morriss-Kay G. Bodyworlds: Exhibition at the Atlantis Gallery London. Journal of Anatomy. 2002;**200**:535-536

[41] Burr DB. Congress considering plastination import ban. American Association of Anatomists News. 2008;**17**:2-5

[42] Rudolph JN, Perlov D. Body Worlds: An anatomical exhibition of real human bodies, Summary of ethical review [Internet]. 2009. Available from: http://www.bodyworlds.com/Downloads/englisch/Media/Press%20Kit/BW_LA_SummaryofEthicalReview.pdf [Accessed: 2016-05-16]

[43] Jones DG. The Peril and Promise of Medical Technology. Oxford: Peter Lang; 2013. p. 280

[44] von Hagens G, Whalley A. Anatomy Art – Fascination Beneath the Surface. Heidelberg: Institute for Plastination; 2000. p. 284

[45] Moore CM, Mackenzie Brown C. Gunther von Hagens and Body Worlds part 2: The anatomist as priest and prophet. The Anatomical Record Part B: The New Anatomist. 2004;**277B**:14-20

[46] Stern M. Dystopian anxieties versus utopian ideals: Medicine from *Frankenstein* to *The Visible Human Project* and *Body Worlds*. Science as Culture. 2006;**15**(1):61-84

[47] Scott R. Body Worlds' plastinates, the human/nonhuman interface, and feminism. Feminist Theory. 2011;**12**(2):165-181

[48] Jespersen TC, Rodríguez A. Forced impregnation and masculinist utopia. In: Jespersen TC, Rodríguez A, Starr J. editors. The Anatomy of Body Worlds: Critical Essays on the Plastinated Cadavers of Gunther von Hagens. Jefferson, NC: McFarland & Co; 2009. pp. 166-175

[49] Hirschauer S. Animated Corpses: Communicating with Post Mortals in an Anatomical Exhibition. Body & Society. 2006;**12**(4):25-52

[50] Linke U. Body Shock: The political aesthetics of death. Social Analysis. 2010;**54**(2):80-98

[51] Linda S-S. Advise and consent: On the Americanization of Body Worlds. BioSocieties. 2006;**1**:369-384

[52] von Hagens G. Gruselleichen, Gestaltplastinate und Bestattungszwang (On gruesome corpses, gestalt plastinates and mandatory interment). In: Wetz FJ, Tag B. Schöne, editors. Neue Korperwelten: Der Streit um die Austellung (Brave New Body Worlds—The Question of the Exhibition); Stuttgart: Klett-Cotta; 2001. pp. 260-282

[53] PRNewswire. Anatomist Dr. Gunther von Hagens Reiterates His Mission of Public Health Education to Press Corps in Guben, Germany. 30 November 2006. Available from: www.prnewswire.co.uk/cgi/news/release?id=185453. [Accessed 2010-09-20]

[54] Jones DG, Whitaker MI. Engaging with plastination and the Body Worlds phenomenon: A cultural and intellectual challenge for anatomists. Clinical Anatomy. 2009;**22**:770-776

[55] Stern M. Shiny happy people: 'Body Worlds' and the commodification of health. Radical Philosophy. 2003;**118**:2-6

[56] Linke U. Touching the corpse: The unmaking of memory in the body museum. Anthropology Today. 2005;**12**(5):13-19

[57] Desmond J. Postmortem exhibitions: Taxidermied animals and plastinated corpses in the theaters of the dead. Configurations. 2010;**16**(3):347-378

[58] Jones DG. The artificial world of plastination: A challenge to religious perspectives on the dead human body. The New Bioethics. 2016;**22**(3):237-252. DOI: 10.1080/20502877.2016.1238580

[59] Nicholls EH. Selling anatomy: The role of the soul. Endeavour. 2002;**26**(2):47

[60] Montross C. Body of Work: Meditations on Mortality from the Human Anatomy Lab. London: Penguin Books; 2007. p. 295

[61] Riederer BM. Plastination and its importance in teaching anatomy. Critical points for long-term preservation of human tissue. Journal of Anatomy. 2014;**224**:309-315

[62] IFAA. Recommendations of good practice for the donation and study of human bodies and tissues for anatomical examination. Plexus (Newsletter of the International Federation of Associations of Anatomists). Available from: http://www.ifaa.net/index.php/plexus/viewdownload/3-plexus/7-plexus-2014 [Accessed 2014-08-20] See also: Ifaa.net [Accessed 2017-02-20]

[63] Jones DG. Searching for good practice recommendations on body donations across diverse cultures. Clinical Anatomy. 2016;**29**:55-59

[64] Jones DG, King MR. Maintaining the anonymity of cadavers: Historic relic or educational and ethical necessity? Anatomical Sciences Education. 2017;**10**:87-97

[65] Champney TH. The business of bodies: Ethical perspectives on for-profit body donation companies. Clinical Anatomy. 2016;**29**:25-29

[66] Bristol Royal Infirmary Inquiry. Learning from Bristol: The Report of the Public Inquiry into Children's Heart Surgery at the Bristol Royal Infirmary 1984-1995. Bristol: Crown Copyright; 2001

[67] Department of Health. The Removal, Retention and Use of Human Organs and Tissue from Postmortem Examination. London: Her Majesty's Stationary Office; 2001

[68] The Royal Liverpool Children's Inquiry. The Royal Liverpool Children's Inquiry Report. London: House of Commons; 2001

[69] Retained Organs Commission. A Consultation Document on Unclaimed and Unidentifiable Organs and Tissue, a Possible Regulatory Framework. London: National Health Service; 2002

[70] Jones DG, Whitaker MI. Speaking for the Dead: The Human Body in Biology and Medicine. Farnham, Surrey, UK: Ashgate Publishing; 2009. p. 296

[71] Jones DG. YouTube anatomy education: Source of ethical perplexity. Anatomical Sciences Education. 2016;**9**:500-501

Chapter 2

Innovative Technologies for Medical Education

Pascal Fallavollita

Additional information is available at the end of the chapter

http://dx.doi.org/10.5772/intechopen.68775

Abstract

This chapter aims to assess the current practices of anatomy education technology and provides future directions for medical education. It begins by presenting a historical synopsis of the current paradigms for anatomy learning followed by listing their limitations. Then, it focuses on several innovative educational technologies, which have been introduced over the past years to enhance the learning. These include E-learning, mobile apps, and mixed reality. The chapter concludes by highlighting future directions and addressing the barriers to fully integrating the technologies in the medical curriculum. As new technologies continue to arise, this process-oriented understanding and outcome-based expectations of educational technology should be embraced. With this view, educational technology should be valued in terms of how well the technological process informs and facilitates learning, and the acquisition and maintenance of clinical expertise.

Keywords: anatomy learning, gross anatomy, mixed reality, medical education, rehabilitation

1. Introduction

Medicine is the science and practice of the diagnosis, treatment, and prevention of disease, and most medical information is about our human body. Medical information is employed in different scenarios, such as education, training, diagnosis, surgery, etc. As the computing technologies develop, more information is digitized from the physical world, and then researchers work on how to process and show them back to the user, enabling the user to perceive and interact with the information naturally and effectively. Perception and interaction with different media and objects are the fundamental human activities, and they are user specific [1].

Anatomical education is an important content of every curriculum and starts already very early in school, to form a good understanding of the body and improve the general population's health awareness [2]. With the advent of the plethora of exciting technological advancements there should be no reason not to include these for the creation of new education paradigms for medical learning. As such, this chapter presents an overview of the fascinating novel research, which is being undertaken in this area.

2. State-of-art

A clinician's knowledge is collected through many years of medical practice. In various settings, the knowledge will be transferred to medical students through medical education and to the general population through public education. For example, in the surgical setting, medical imaging data are collected for diagnosis and navigation. Good communication between the patient and the surgeon is very important for making the patient comply with the surgical procedure. The same communication plays a vital role for patients who perform rehabilitation exercises after surgery. In all of these scenarios, key players act different roles to perceive and interact with medical information. In this section, we discuss the state-of-art and challenges involved in medical education, public education, and the communication to prepare for surgery and rehabilitation.

2.1. Medical education

Though medical education has been existent for centuries, we begin from Abraham Flexner's report in 1910 [3]. Flexner visited 155 medical schools that existed in North America in 1909, and he identified four major problems during medical education: (1) lack of standardization, (2) lack of integration, (3) lack of inquiry, and (4) identity formation. The report had a huge impact and shaped modern medical education, where patient care, teaching, and research are combined. However, academic hospitals do not spend enough time on teaching due to enormous pressure to publish, as well as, for economic reasons [4]. Another issue is that research is focusing on very small subtopics, which do not relate to medical education [5]. On the other hand, only little research is done for teaching. An example is gross anatomy where most topics are already known and only little novel research exists. In today's medical schools students are required to understand both functional and spatial context of human anatomy.

Traditional methodologies: Traditional medical education learning is classified into three categories: cadaver, model, and textbook. Although technology has advanced significantly in the last decades, school education still mostly uses the same methods to convey anatomical knowledge [2]. Typically, the information is collected in printed books like anatomy atlases, displayed in the form of charts and diagrams. Those diagrams provide a simple and well-known method to illustrate form and appearance of organs, having the advantage that the user is accustomed to such methods of display. However, there exist several downsides to this method. First of all, the view is limited to a selected few different cross-sections the author chose to present. This may not be sufficient in some cases to fully convey how an organ is

http://dx.doi.org/10.5772/intechopen.68775

located relative to its surroundings since occlusions limit the possibilities to visualize these spatial relations [6]. Another problem is that often the organs are only depicted schematically by leaving out details or distorting tissue colors; thus, giving only a coarse impression on how the organ actually looks like in reality. For example, it is difficult to interpret the spatial and physical characteristics of anatomy by observing two-dimensional images, diagrams, or photographs. Many physical models also lack detail levels to fully understand the specific anatomy (see **Figure 1**).

Anatomy education is also performed by the dissection of cadavers. The value of dissection classes as a teaching format lies in the fact that it provides a 3D view on human anatomy including tactile learning experiences. It enables elaboration of knowledge already acquired in lectures and study books and it provides an overall perspective of anatomical structures and their mutual relations in a whole organism [7]. The cadaver-based learning has seen decline due to practical and cost issues (see **Figure 2**). And so far, no objective empirical evidence exists concerning the effectiveness of dissection classes for learning anatomy [8].

Computer-based learning: It is developed by experts and students can use these materials if there is no available expert in the hospital. Computer-based education can be very powerful for anatomy teaching, where 3D visualization is of great benefit (see **Figure 3**). Many virtual model databases exist and E-learning is commonly used today. These resources are valuable and are more interactive and interesting than textbooks. In addition, the personalization in E-learning systems has been the subject of much recent research and allows teachers to select parameters and combine them flexibly to define different personalized strategies according to the specifics of the courses. In recent times, digital approaches are evolving, which are trying to improve on these methods. Many of them offer interactive anatomy models usable either as an on-line service or as a standalone application. There also exist organizations specifically offering teaching bundles for use in classrooms, making it possible to use alternative teaching methods at different school levels assisted by videos and interactive tools.

Those interactive applications offer large improvements over the traditional method. It is possible to view organs and structures from any desired angle, control magnification and often even select specific organs and systems to be displayed or hidden. Many commercially available

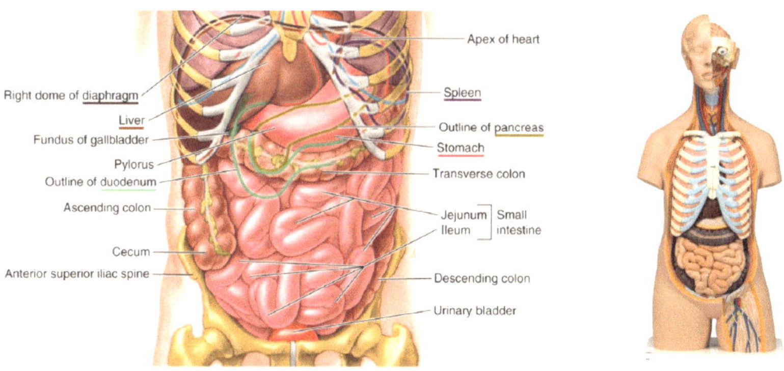

Figure 1. Traditional Atlas and physical model methodologies for medical learning.

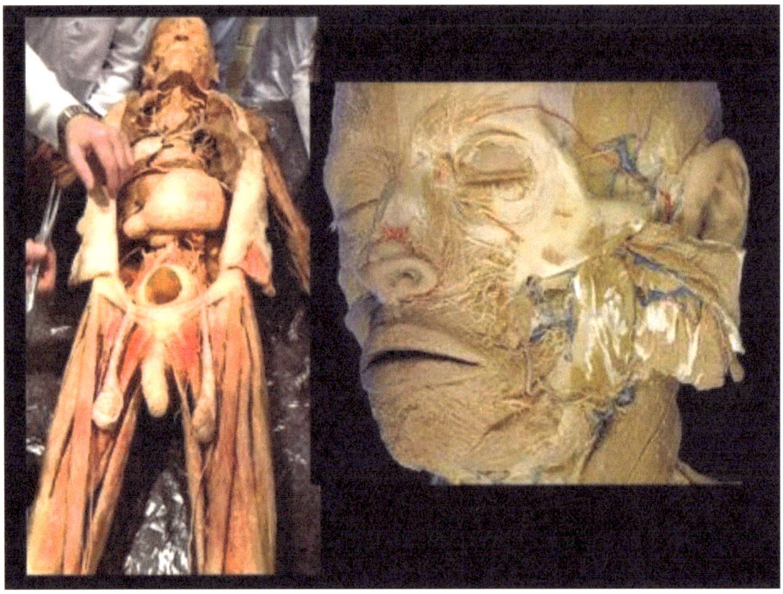

Figure 2. Anatomy learning with cadavers.

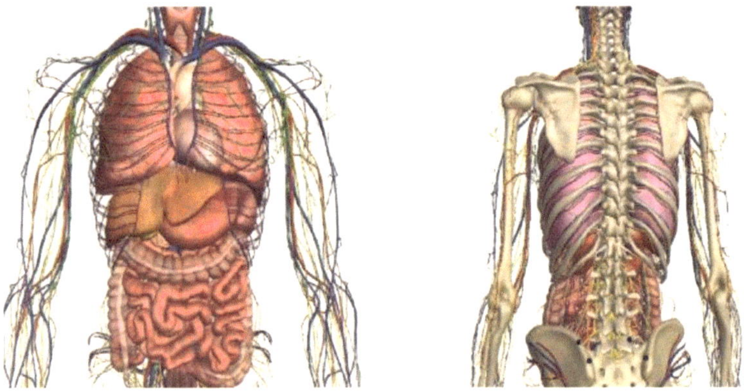

Figure 3. Anatomy 3D models.

systems use virtual reality for medical education and psychomotor skills training [9, 10]. These systems have indeed proven to be valid and useful. By combining computer models of anatomical structures with custom software, we can demonstrate to students the new ways of interacting with anatomy that could not be achieved during cadaveric dissections or in static diagrams and models, thereby increasing their learning satisfaction [11]. Visualization of medical data can also be used for education. Medical datasets are very large and the visualization must correctly represent the reality. Today, images with a high resolution can be generated by performance graphics for computer gaming (see **Figure 4**). Through all those materials and interactive elements, the users can control the region of interest and hopefully better understand the information they are

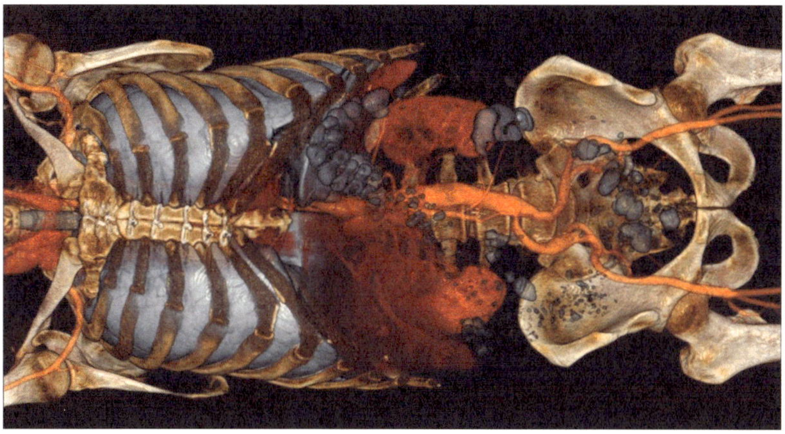

Figure 4. Recent CT volume visualization.

looking for. These resources are very valued, however, with the disadvantage of having users still mentally map/link anatomy onto their bodies. Humans usually have to build the knowledge network in the brain after perception of the knowledge. The information is very general but the learning procedure is quite personal. Personalized information is much easier for the user to accept. Novice medical students accept a very long and difficult study during medical school. At the same time, different technologies are developed to simplify the learning process.

2.2. Public and patient education

Methodologies for education in medical schools are not suitable for public education, as it is not necessary to teach a high level of anatomy understanding. Real-life demonstrations are usually limited to the skin surface, which is why most people learn about anatomy in the form of charts or plastic models of the inner organs. Still, there is a certain fascination about looking inside one's own body, something that can usually only be achieved to a limited degree by the use of X-ray imaging or similar imaging modalities. These methods are not applicable for public education, mostly because the devices are too expensive to use without medical indication. Patient education is the process by which health professionals impart information to patients and their caregivers that will alter their health behaviors or improve their health status. It can improve the patient's understanding of medical condition, diagnosis, disease, or disability. Patient education is an important and often underestimated responsibility of a health professional [12]. It is the responsibility of a doctor to inform and motivate patients to ensure that they understand the diseases, the treatment options available and the consequences of no treatment or noncompliance [13]. This makes it possible for the patients to understand it, eventually allowing them to take responsibility for their own health [14]. Good information and communication increase the patient's possibility to contribute in the decision-making process, leading to higher levels of patient satisfaction, loyalty to the physician and favoring treatment outcomes [14–16]. Traditionally, patient-health professional communication in a clinical setting comprises of a face-to-face narrative interaction often combined with static images or real-time sketching (see **Figure 5**-left). It is inevitable

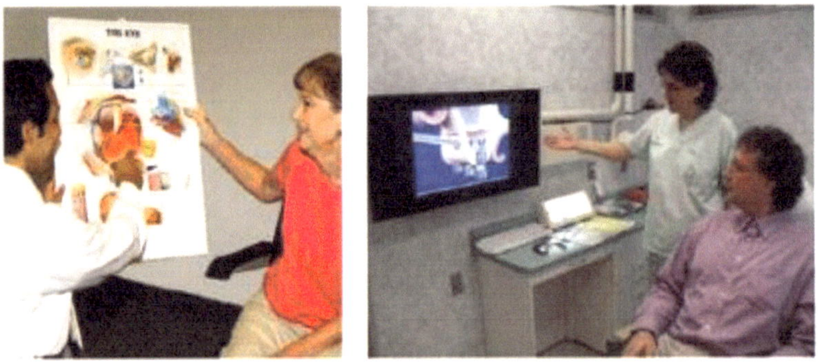

Figure 5. Patient education scenarios using Atlas or multimedia.

that patients' comprehension of explanations is often lacking due to the complexity of the medical information [16]. Evidence suggests that patients often do not understand what is being said to them when information is given during a medical encounter due to cultural and educational gaps between clinicians and patients [17]. This discourages the patient and leaves them overly reliant on their doctor's advice. Improving and facilitating the information process and understanding by the patient has been a focus of research in the medical field. Ong et al. [18] presented a literature survey of doctor-patient communication, summarizing the key topics of information exchange, interpersonal relationship building, and shared medical decision making. Wilcox et al. [19] proposed a design for patient-centric information displays to deliver useful information to a patient during an emergency room visit, since the patients are frequently under informed. Previous work has also demonstrated that technology can positively impact communication. Hsu et al. [20] conducted a longitudinal study the impact of in-room access to electronic health records on doctor-patient interactions during outpatient visits. Recent reviews show growing evidence with regard to the benefit of multimedia tools in enhancement of patients' satisfaction and improvement of knowledge retention [17]. A patient education computer program using 3D multimedia videos is shown in **Figure 5**-right.

In addition, 3D animations have been suggested, for educating audiences that are preliterate or have limited literacy skills, such as children with mental handicaps [21] and to be useful for patients with a lower learning pace as it leads to a better understanding of the disease [22]. Ni et al. [23] presented the design, development, and evaluation of AnatOnMe, a projection-based handheld device designed to facilitate medical information exchange (see **Figure 6**). Adopting a user-centered design approach, they interviewed medical professionals to understand their practices, attitudes, and difficulties in exchanging information with patients, as well as to identify their workflow, tasks, and design requirements for a technology intervention. Gonzales and Riek [24] presented an application that personalized the content presented on a device of the patient's diagnosis in an easy-to-understand language, rather than hard to understand medical terminology. It also encourages doctor-patient interaction on the main relevant areas of the patient's diagnosis. Ihrig et al. [25] evaluated the view of physicians performing multimedia supported preoperative education within a randomized controlled trial, enrolling 8 physicians and 203 patients. Both patients and physicians profited from multimedia support for education and counseling without prolonging patient education.

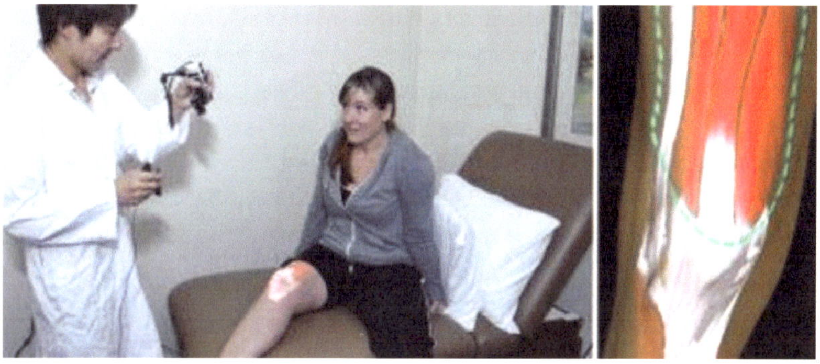

Figure 6. AnatOnMe in use in a simulated physical therapy consultation.

2.3. Rehabilitation exercises

Rehabilitation is the process to regain full function following injury. This involves restoring strength, flexibility, endurance and power and is achieved through various exercises and drills (see **Figure 7**). Rehabilitation is as important as treatment following an injury but unfortunately is often overlooked. Usually physiatrists are involved in the exercise procedure and optimal physical activities are carried out in an effort to reach specific health objectives. Its purpose is to return to normal musculoskeletal function or to decrease pain caused by injuries. The communication between a medical assistant and a patient is very important during the exercises, e.g., make sure that the patient knows the current state of his/her disability and follows guidelines during the rehabilitation process. The rehabilitation exercises are commonly performed in a rehabilitation center, and the physiotherapist identifies and evaluates the movements and motor functions that are being affected. In order to achieve kinetic gains as quickly as possible, patients are encouraged to perform the same exercise movements several times [26]. Patients' learning by movement repetition is crucial for their successful therapy.

The time that the patient expends in the rehabilitation center is relatively short compared with the time he/she spends at home with no supervision [27]. However, there are some issues with home exercises: (i) the patient cannot be motivated by the therapist and (ii) wrong movements might be performed without timely correction. Today, interactive solutions for rehabilitation

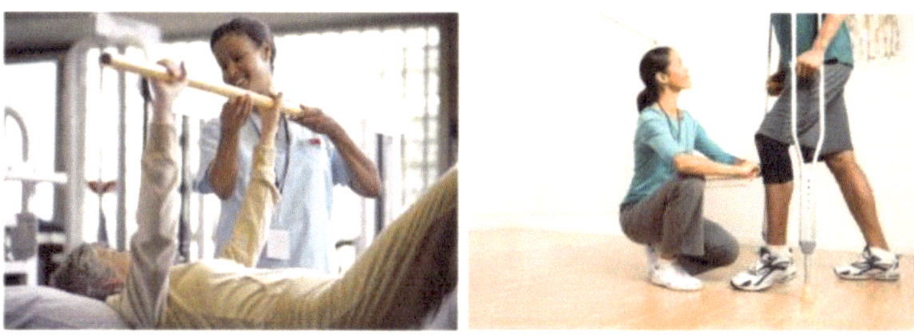

Figure 7. Physiotherapists leading the exercises during rehabilitation therapy.

are being developed using different sensors and motion tracking systems including gloves, magnetic, fiducial or infrared markers, video and depth cameras. Mixed-reality systems are also introduced to motivate the patient and facilitate a more accurate exercise [28, 29]. The activity of patients' muscles can be visualized and the movement is checked based on the contracted muscles [30]. Cho et al. [29] developed a virtual-reality proprioception rehabilitation system for stroke patients to use proprioception feedback in upper limb rehabilitation by blocking visual feedback. The markerless tracking feature of Kinect enables a natural user interaction for rehabilitation applications, which alleviates existing issues for patients having difficulty to hold any sensor or marker. Using Kinect, various researches are being explored for the development of assistive systems that help interact with patients during their rehabilitation exercises [31–34].

3. The magic mirror

Providing adequate learning experience to different learners is a challenging issue as the traditional techniques generally cannot adapt content to suit the individual learner's needs. Personalization for promoting a multi-modal learning environment is a growing area of interest, such as the development of user modeling and personalized processes, which place the student at the center of the learning development. Augmented reality systems can present a virtual representation of the subject material and create a direct connection between the information the user wants to learn and their own body, and activities at the same time. Hence, it could help understand and memorize complex information, and either supplement conventional learning or even supersede it altogether. Previous mixed-reality systems on visualization of anatomy used expensive technologies involving head mounted displays (HMDs) or markers. Recently, a Magic Mirror technology has been developed for anatomy education, by employing a camera for tracking and a TV display for visualization to of an augmented reality view [35]. The system presented is both inexpensive and easy to use. It presents medical data augmented onto the user's body and shows additional 2D and 3D information according to the user's needs. The Magic Mirror provides the user 'superman ability' to look into their body. It enables the medical information to be perceived naturally linking it to a real human body. Natural gesture is chosen as the interaction methodology, and interaction with the augmented reality view of the user's body provides a personalized perception in the Magic Mirror framework.

The Magic Mirror [36] is a user interface technique that mimics a normal mirror and presents nonphysical visual feedback in addition to the normally optical effect. Here the user stands in front of a screen and via a camera, the image of the user is shown on the screen such that it acts like a mirror. While previous systems have augmented objects onto the user, this system extends the concept for medical education and rehabilitation. A depiction of this can be seen in **Figure 8**.

To achieve this visualization, the Magic Mirror augments the volume visualization of a CT dataset onto the user. To allow a correct augmentation of the medical data, the pose of the user is tracked. The Magic Mirror concept came out as a framework and it generates a personalized perception of the medical information for every user. In addition, the framework takes the user's natural gesture as input to create an interactive mixed-reality environment.

http://dx.doi.org/10.5772/intechopen.68775

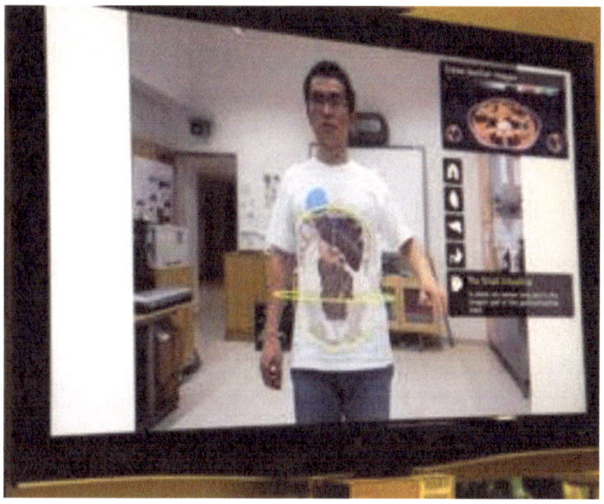

Figure 8. By using the Magic Mirror system, the user is led to believe that he or she is able to look inside their own body.

Knowledge about human anatomy is an important issue for everyone working in the field of medicine. It is also an important part of the general education and relevant for many other professions related, e.g., to health-care or sports. Human anatomy is very complex and it does not only involve knowledge about a single organ, but also about issues such as chemical processes, human motion and spatial relations inside the body. Consequently, teaching human anatomy is very difficult and often a large effort is spent on teaching it; e.g., by letting students perform dissection courses, creating illustrations, using plastic models of anatomy or by utilizing 3D computer graphics. The Magic Mirror framework is firstly employed to display anatomical structures overlaid onto the body of the user to intuitively teach human anatomy.

3.1. Hardware setup

The Magic Mirror technology focuses on a few important organs of the abdomen, namely the liver, lungs, pancreas, stomach, and bones. The system prototype has a mirror-like effect to the user by projecting a 'looking glass' on the body and displays the skeleton of the user, rendered from CT data and anatomy 3D models. The framework tracks users' movements using a depth camera and an algorithm to detect the pose of the user from the depth image. This is realized using the Microsoft Kinect, which was originally developed to allow controlling computer games by motion. By using the Magic Mirror metaphor, the user is led to believe that he or she is able to look inside their own body. At the same time, medical information (CT, MRI data, and a fully segmented dataset of cross-sectional photographs of the human body) is augmented in real-time. The current system also allows visualization of static anatomy on the user and offers a simple user interface to select CT, MRI, or photographic slices [35, 37]. An illustration of the hardware setup can be seen in **Figure 9**. The first component of the system is a display device. In different setups of the technology, large TV screens or video projection onto a planar surface has been used. The second component is a color camera, which is mounted next to the display surface and which is looking at the user and enables visual perception of the information.

Figure 9. Magic Mirror hardware setup showing the general design and the current implementation using Kinect.

The third component is a depth sensor, which is placed next to the color camera and which has a similar field of view and viewing direction as the color camera to collect the user's skeleton information. The current system uses the Microsoft Kinect V1, which consists of a color and a depth camera that are assembled into a mechanical housing. The depth sensor is an infrared camera that uses structured light, which is emitted by an additional infrared projector to estimate depth values for each pixel. The user's skeleton and personal information can be generated from the Kinect sensor based on machine learning. The system employs the color camera to create a mirror-like effect to the user, and all the nonphysical visual feedbacks are generated based on the user's skeleton and personal information via volume rendering the corresponding medical information onto the human body.

3.2. Software setup

The system framework of the prototype is illustrated in **Figure 10**. To access the Kinect the system uses Microsoft Kinect SDK or OpenNI1, which is an open source software framework

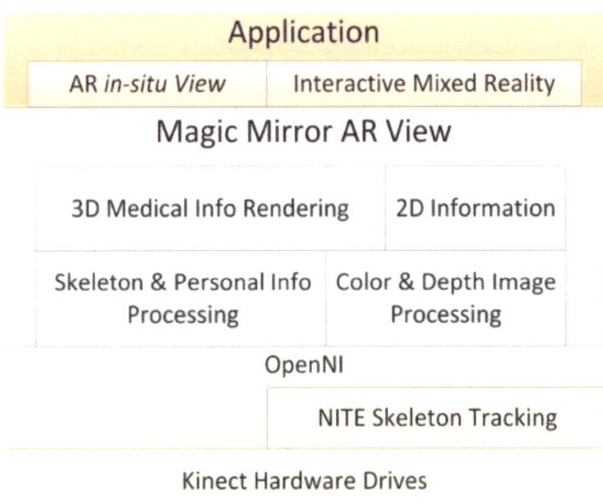

Figure 10. Magic Mirror system framework. The lowest two layers represent the open source library to access the Kinect raw image data and skeleton information. The middle layer is the modules used to create Magic Mirror AR view of the user. The top level is the application with basic system features.

that allows retrieving color and depth images from the Kinect. The depth sensor is used for two purposes. First, the depth values are projected to the color image providing depth information for each pixel in the color image. Second, a skeleton tracking algorithm uses the depth image to track the pose of multiple joints of a user who is standing in front of the camera. For skeleton tracking, the Magic Mirror uses NITE, software by PrimeSense, that performs gesture recognition and skeleton tracking based on depth images. NITE can be used with the Kinect through the OpenNI framework. The Magic Mirror augmented reality view is based on the information perceived via Kinect and corresponding medical information.

As shown in **Figure 10**, there are four modules: Skeleton & personal information processing is important to achieve personalized perception. Color & depth image processing is the basic module to generate the mirror-like effect and merge the nonphysical visual outcome. 3D medical information rendering processes the corresponding 3D medical images or models and generates virtual elements for the Magic Mirror augmented reality view and it can employ OpenGL and OpenGl Shading Language, and any other 3D rendering libraries, e.g., Coin3D. 2D information includes window management and basic user interface elements, such as 2D text and image information, and it can be implemented using Qt. The applications based on the Magic Mirror AR View have two important features, AR *in situ* view and interactive mixed reality. The blue rectangles represent the basic visualization functions of the system.

3.3. Medical information

Aside from text and 2D atlases, 3D volumes and models are also important methodologies to represent medical information. The common way in medicine is to use 3D volumes, where we have one pixel or voxel for every point in 3D. This is the kind of data we get from CT or MRI. The other ones are polygonal or surface models, where only the surface (e.g., skin or the surface of the organ) is stored. For models that have been created for education often surface models are used because they look better. While the Magic Mirror technology can use a CT or MRI scan, it does not make sense to acquire a CT or MRI scan of the user if it is not required for medical reasons. Therefore, we augment the medical volume from the visible Korean human (VKH) dataset onto the user [38]. This dataset consists of a CT scan, an MRI volume and a photographic volume, which has been acquired by stacking up cryosections. Most CT and MRI medical images are saved in the DICOM standard, which is the format that is used in all hospitals. Unfortunately, most software for research does not support DICOM. The Magic Mirror system takes an *.MHD file* as the medical data. One drawback of the volume data is that the 3D dataset cannot be deformed in real time. So if the user bends, this is not reflected in the visualization of the medical data leading to movements of the limbs not visualized correctly. Possible solutions to address this issue have been published in Section 3.4, but for the current technology, which focuses on the abdominal area, this is a minor issue. Visualizing structures other than bones from the CT is more challenging. In a first attempt, the segmentation that is available for the CT volume was used to visualize different organs in the abdominal area. The quality of the visualization was low, as the segmentation does not have subpixel accuracy and transfer functions on CT intensities cannot provide visualization with realistic colors and textures of organs. Therefore instead of using the volumetric data, additional polygonal models were integrated. The Anatomium2 dataset provides polygonal models of many organs of the human body. A scene graph including multiple organs was extracted from the dataset. Using Coin3D this scene graph is augmented onto the user. The simultaneous visualization of bones from CT and a polygonal model of the small intestine is shown in **Figure 11**.

3.4. Related publications

It is my firm belief that providing a more personalized solution allowing *in situ* contextual visualization onto one's body would provide a greater learning experience when compared to existing solutions. Not only does 'seeing inside the body' spark curiosity and excitement, it also permits a sense of ownership and fidelity during the learning process. The Magic Mirror technology allows visualization of medical and radiological CT data directly on the user body and a gesture-based user interface (UI) allows direct interaction with this data [39]. A large user study was carried out earlier in 2016 to verify the learning potential and acceptability of the technology. The Magic Mirror was tested and evaluated by 748 first and second semester medical students at the Anatomy Department at the Klinikum der Universität München, Germany, and the learning outcomes were extremely positive, particularly with respect to three dimensional understanding of organs and better comprehension of anatomical structures. Preliminary work in rehabilitation research [40, 41] resulted in two publications in 2016. The first introduced a MirrARbilitation system that improves patient exercise engagement and performance quality. Compared to state-of-art, the system consists of a gesture recognition tool based on the International Biomechanical Standards terminology for the reporting of human joint motion. An improvement from 69.02 to 93.73% of correct exercises on a cohort of 33 end users was demonstrated by the system. The second introduced a mixed-reality system facilitating the learning of the muscles of the upper extremities. The system consists of two main components: an augmented reality view superimposing the virtual model of the arm on top of the video stream and a virtual reality view, providing a more detailed image of the

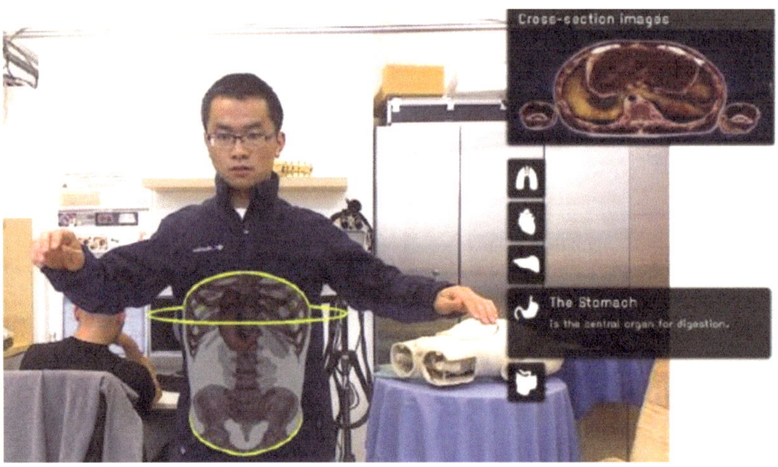

Figure 11. The AR *in situ* visualization of human anatomy in the Magic Mirror technology.

http://dx.doi.org/10.5772/intechopen.68775

muscles. A user study including 20 students was performed indicating that the system is useful for learning the structure and function of the muscle and can be a valuable supplementary to established muscle learning paradigms.

4. Conclusion

This chapter presented solutions to educate the medical and general population. It concluded by highlighting a personalized Magic Mirror system for anatomy and rehabilitation education. The system combines both augmented and virtual reality environments, which are certified useful for anatomy learning. Both academic and medical center methodologies must be changed according to the advances of new technologies, but there is still a long way for this. Considering the benefits of the personalized and interactive systems for motivation and perception of anatomy learning, new technologies can additionally be helpful to facilitate autonomous learning and secondarily to reduce laboratory material and instructor costs. Together with the anatomy & medical community, we hope to initiate such discussions in integrating exciting user-specific and gaming concepts via new anatomy learning systems and ultimately for improved patient education.

Acknowledgements

I would like to thank the Chair for Computer Aided Medical Procedures & Augmented Reality, Technische Universität Munchen, for allowing me the great opportunity of growing my research in this area.

Author details

Pascal Fallavollita

Address all correspondence to: pfallavo@uottawa.ca

Interdisciplinary School of Health Sciences, University of Ottawa, Canada

References

[1] Tan HZ. Perceptual user interfaces: Haptic interfaces. Communications of the ACM. 2000;**43**(3):40-41

[2] Dunnill R. Education—An anatomy of the discipline: Rescuing the university project? Widening Participation and Lifelong Learning. 2013;**15**(4):92-95

[3] Flexner A. Medical education in the United States and Canada. Bulletin of the World Health Organization. 2002;**80**(7):594-602.

[4] Saidi F. American medical education 100 years after the Flexner report. Archives of Iranian Medicine. 2007;**10**(1):131.

[5] Ludmerer KM. The internal challenges to medical education. Transactions of the American Clinical and Climatological Association. 2003;114-241

[6] Gilroy A. M, MacPherson B. R, Ross L. M, Schünke M, Schulte E, & Schumacher, U. (2012). Atlas of anatomy. New York: Thieme. p. 704

[7] McLachlan JC, Bligh J, Bradley P, Searle J. Teaching anatomy without cadavers. Medical Education. 2004;**38**(4):418-424

[8] Frank JR. (Ed). 2005. The CanMEDS 2005 physician competency framework: Better standards. Better physicians. Better care. Ottawa: The Royal College of Physicians and Surgeons of Canada.

[9] Lu J, Pan Z, Lin H, Zhang M, Shi J. Virtual learning environment for medical education based on VRML and VTK. Computers & Graphics. 2005;**29**(2):283-288

[10] Basdogan C, Sedef M, Harders M, Wesarg S. VR-based simulators for training in minimally invasive surgery. IEEE Computer Graphics and Applications. 2007;**27**(2):54-66

[11] Bacca J, Baldiris S, Fabregat R, Graf S. Augmented reality trends in education: A systematic review of research and applications. Journal of Educational Technology & Society. 2014;**17**(4):133

[12] Waitzkin H. Doctor-patient communication: Clinical implications of social scientific research. JAMA. 1984;**252**(17):2441-2446

[13] Fathi A, Farhadi A, Rehg JM. Understanding egocentric activities. In: International Conference on Computer Vision (ICCV); 2011. pp. 407-414

[14] Ammanna A, Kissb T, Klebbac Ä, Matthiesd HK. The next generation of patient education: Multilingual Dental Explorer 3D. International Journal of Computerized Dentistry. 2010;**13**(1):43-55

[15] Huber J, Ihrig A, Herzog W, Huber CG, Konyango B, Löser E, Hatiboglu G, Hadaschik BA, Pahernik S, Hohenfellner M. Patients' view of their preoperative education for radical prostatectomy: Does it change after surgery?. Journal of Cancer Education. 2012;**27**(2):377-382

[16] Cleeren G, Quirynen M, Ozcelik O, Teughels W. Role of 3D animation in periodontal patient education: A randomized controlled trial. Journal of Clinical Periodontology. 2014;**41**(1):38-45

[17] Beranova E, Sykes C. A systematic review of computer-based softwares for educating patients with coronary heart disease. Patient Education and Counseling. 2007;**66**(1):21-28

[18] Ong LM, De Haes JC, Hoos AM, Lammes FB. Doctor-patient communication: A review of the literature. Social Science & Medicine. 1995;**40**(7):903-918

[19] Wilcox L, Morris D, Tan D, Gatewood J. Designing patient-centric information displays for hospitals. In: Proceedings of the SIGCHI Conference on Human Factors in Computing Systems; 2010. pp. 2123-2132

[20] Hsu J, Huang J, Fung V, Robertson N, Jimison H, Frankel R. Health information technology and physician-patient interactions: Impact of computers on communication during outpatient primary care visits. Journal of the American Medical Informatics Association. 2005;**12**(4):474-480

[21] Palin-Palokas T, Nordblad A, Remes-Lyly T. Video as a medium of oral health education for children with mental handicaps. Special Care in Dentistry. 1997;**17**(6):211-214

[22] Jimison HB, Sher PP, Appleyard R, LeVernois Y. The use of multimedia in the informed consent process. Journal of the American Medical Informatics Association. 1998;**5**(3):245-256

[23] Ni T, Karlson AK, Wigdor D. AnatOnMe: facilitating doctor-patient communication using a projection-based handheld device. In: Proceedings of the SIGCHI Conference on Human Factors in Computing Systems; 2011. pp. 3333-3342

[24] Gonzales M, Riek L. A shared interface to improve oncologist-patient communication. In: Proceedings of the 6th International Conference on Pervasive Computing Technologies for Healthcare; 2012. p. 14

[25] Ihrig A, Herzog W, Huber CG, Hadaschik B, Pahernik S, Hohenfellner M, Huber J. Multimedia support in preoperative patient education for radical prostatectomy: The physicians' point of view. Patient Education and Counseling. 2012;**87**(2):239-242

[26] Metcalf CD, Robinson R, Malpass AJ, Bogle TP, Dell TA, Harris C, Demain SH. Markerless motion capture and measurement of hand kinematics: Validation and application to home-based upper limb rehabilitation. IEEE Transactions on Biomedical Engineering. 2013;**60**(8):2184-2192

[27] Crocher V, Hur P, Seo NJ. Low-cost virtual rehabilitation games: House of quality to meet patient expectations. In: International Conference on Virtual Rehabilitation (ICVR); 2013. pp. 94-100

[28] Da Gama A, Fallavollita P, Teichrieb V, Navab N. Motor rehabilitation using Kinect: A systematic review. Games for Health Journal. 2015;**4**(2):123-135

[29] Cho S, Ku J, Cho YK, Kim IY, Kang YJ, Jang DP, Kim SI. Development of virtual reality proprioceptive rehabilitation system for stroke patients. Computer Methods and Programs in Biomedicine. 2014;**113**(1):258-265

[30] Woodford HJ, Price CIM. EMG biofeedback for the recovery of motor function after stroke. Cochrane Database of Systematic Reviews 2007, Issue 2. Art. No.: CD004585. DOI: 10.1002/14651858.CD004585.pub2.

[31] Cordella F, Di Corato F, Zollo L, Siciliano B, van der Smagt P. Patient performance evaluation using Kinect and Monte Carlo-based finger tracking. In: International Conference on Biomedical Robotics and Biomechatronics (BioRob); 2012. pp. 1967-1972

[32] Da Gama A, Chaves T, Figueiredo L, Teichrieb V. Guidance and movement correction based on therapeutics movements for motor rehabilitation support systems. In: Virtual and Augmented Reality (SVR); 2012. pp. 191-200

[33] Antón D, Goñi A, Illarramendi A, Torres-Unda JJ, Seco J. KiReS: A Kinect-based telerehabilitation system. In: e-Health Networking, Applications & Services (Healthcom), editors. 2013. pp. 444-448

[34] Borghese NA, Mainetti R, Pirovano M, Lanzi PL. An intelligent game engine for the at-home rehabilitation of stroke patients. In: Serious Games and Applications for Health (SeGAH); 2013. pp. 1-8

[35] Blum T, Kleeberger V, Bichlmeier C, Navab N. mirracle: Augmented Reality *in-situ* visualization of human anatomy using a magic mirror. In: Virtual Reality Short Papers and Posters (VRW); 2012. pp. 169-170

[36] Grosjean J, Coquillart S. The magic mirror: A metaphor for assisting the exploration of virtual worlds. In: Spring Conference on Computer Graphics; 1999. pp. 125-129

[37] Navab N, Blum T, Wang L, Okur A, Wendler T. First deployments of augmented reality in operating rooms. Computer. 2012;**45**(7):48-55

[38] Park JS, Chung MS, Hwang SB, Shin BS, Park HS. Visible Korean human: Its techniques and applications. Clinical Anatomy. 2006;**19**(3):216-224

[39] Ma M, Fallavollita P, Seelbach I, Heide AM, Euler E, Waschke J, Navab N. Personalized augmented reality for anatomy education. Clinical Anatomy 2015;**29**:446-453

[40] Da Gama AE, Chaves TM, Figueiredo LS, Baltar A, Meng M, Navab N, Teichrieb V, Fallavollita P. MirrARbilitation: A clinically-related gesture recognition interactive tool for an AR rehabilitation system. Computer Methods and Programs in Biomedicine 2016;**135**:105-114

[41] Ma M, Jutzi P, Bork F, Seelbach I, von der Heide AM, Navab N, Fallavollita P. Interactive mixed reality for muscle structure and function learning. In: International Conference on Medical Imaging and Virtual Reality; 2016. pp. 117-128

Chapter 3

Anatomical Correlation of the Arterial Blood Supply to the Spinal Cord in Human and Experimental Animals: A Review

Slavka Flesarova and David Mazensky

Additional information is available at the end of the chapter

http://dx.doi.org/10.5772/intechopen.69119

Abstract

Several animal models exist to examine physiological and functional changes after the spinal cord injury with aim to explain knowledge about the spinal cord injury in human. Before the appropriate animal model is chosen, many aspects must be considered to eliminate the wrong interpretation of the results. The knowledge of the arterial blood supply to the spinal cord is very important in planning the procedures of the spinal cord treatment as well as in animal experiments. As the literature on the topic is disarranged, the aim of this review is to summarize the available literature into one coherent format. This chapter compares the arterial spinal cord blood supply of the frequently used species (pig, dog, cat, rabbit and rat) in experimental spinal cord injury and in human. A complete understanding of the anatomy of the arterial blood supply to the spinal cord is critical for the anatomists and clinicians to determinate the advantages and disadvantages of each animal model for next studies.

Keywords: anatomy, arterial blood supply, experimental animal, human anatomy, spinal cord injury

1. Introduction

Various pathological conditions, including surgical treatments, traumatic injuries, embolism, malformations and tumors, result in severe changes in the arterial blood supply to the spinal cord [1]. The degree and type of present symptoms depend on the affected part of the spinal cord. Spinal cord injury is associated with sustainable disability and results in loss of bladder, respiratory, cardiac, or sexual functions, and in varying degree of paralysis [2]. The superficially located fine arterial system of the spinal cord predisposes the induction of the spinal cord ischemia by spinal cord injuries.

The arterial blood supply to the spinal cord is most profusely documented in human [1, 3, 4]. Several animal species serve in different experimental studies of spinal cord injury as models: pig [5–7], dog [8–10], cat [11–13], rabbit [14–16], guinea pig [17–19], rat [20–22] and mouse [23–25]. To understand the physiologic and pathophysiologic mechanisms of various spinal cord diseases and the effects of several neuroprotective drugs, the detailed knowledge of the spinal cord arterial blood supply is necessary. The following review discusses the current knowledge, principles, peculiarities, variations and known differences in the arterial blood supply to the spinal cord in the most frequently used species of experimental animals and in human.

2. Pattern of the arterial blood supply to the spinal cord

The spinal cord is mainly supplied by the segmental spinal branches. The majority of them originate in the cervical part from the vertebral arteries, in the thoracic part from the dorsal intercostal arteries and in the lumbar part from the lumbar arteries.

2.1. Cervical part of the spinal cord

The cervical part of the spinal cord is except the spinal branches arising from the vertebral artery also supplied by numerous small branches originating from the posterior inferior cerebellar artery. They supply the most cranial part of the first cervical segment of the spinal cord.

The vertebral artery belongs to the branches of the subclavian artery. Through the thoracic inlet leaves the vertebral artery the thoracic cavity. After it's leaving of the thoracic cavity continues towards the transverse foramen of the sixth cervical vertebra in the craniodorsal direction. Inside the transverse canal of the cervical vertebrae, it continues cranially. The vertebral artery reaches the lateral foramen of the atlas through which it enters the vertebral canal. On the floor of the vertebral canal, both vertebral arteries unite together to form the basilar artery, which continues cranially and connects to the arterial circle of the brain. The segmental spinal branches are arising from the vertebral artery along its course inside the transverse canal. They enter the vertebral canal at each segment through the lateral vertebral or intervertebral foramina.

2.2. Thoracolumbar part of the spinal cord

The thoracolumbar part of the spinal cord receives the arterial blood from the spinal branches originating from the paired segmental arteries: the dorsal intercostal and lumbar arteries.

The paired dorsal intercostal arteries originate from the thoracic aorta and/or from some branches of the subclavian artery. The number of dorsal intercostal arteries and the number of the intercostal spaces is the same. The last dorsal intercostal artery is designated as the dorsal costoabdominal artery. They convey the arterial blood to the caudal cervical segments and thoracic part of the spinal cord.

Each dorsal intercostal artery is directed towards the corresponding vertebral body. After it crosses the vertebral body, it gives off the dorsal branch which runs caudoventrally to the transverse process. The main continuation of the artery is directed to the corresponding intercostal space. After the passage of the intercostal space, the interspinal branch to the muscles, interspinal ligaments and spinous processes arise. The spinal branch originates at the level of the intervertebral foramen [26].

The segmental paired lumbar arteries give off the spinal branches supplying the lumbar part of the spinal cord. In almost all cases, the lumbar arteries arise from the dorsal surface of the abdominal aorta.

Each lumbar artery runs on the lateral surface of the corresponding lumbar vertebral body, is directed to the caudal margin of the transverse process and thereafter gives off a branch designed for the vertebral body. The spinal branch arises from the lumbar artery as the second branch. The main continuation of each lumbar artery follows the transverse process of the corresponding lumbar vertebra [26].

2.3. Spinal branches

The majority of the spinal branches supplying the spinal cord originate from the vertebral arteries, dorsal intercostal arteries, and lumbar arteries. At each segment, they enter the vertebral canal passing through the intervertebral foramen. Their passage through the foramen is associated with the corresponding nerve root where they are superficially covered by the perineurium. The spinal branches are loosely attached to the nerve root inside the subarachnoidal cavity. Each of the spinal branches divides into the dorsal or ventral branch, or in some cases in both of them.

The spinal branches are divided into three types:

1. The spinal braches supplying the nerve root or dura mater. They do not reach the spinal cord.
2. The spinal branches ending in the superficial arterial system of the spinal cord.
3. The spinal branches reaching and supplying the spinal cord.

The spinal branches convey the arterial blood to the corresponding nerve root, spinal ganglion, dura mater, dorsal spinal arteries and ventral spinal artery [27]. The calibre of the spinal branch determines which structures are supplied. The spinal branch with smaller diameter very often supplies only the nerve root [28].

On the dorsal and ventral surface of the spinal cord are present the dorsal and ventral branches arising from the spinal branches divided into the small cranial and caudal branches. In some cases, these branches are recognizable and their fusion constitutes the dorsal spinal arteries and the ventral spinal artery.

The dorsal and ventral branches occurring at the same level and side, give off small superficial branches which constitute the spinal arterial ring [28, 29] or the *vasa coronae* [30]. The pial arterial plexus is formed by a network of anastomoses encircling the spinal cord. The majority of branches arising from the pial arterial plexus enter the midline of the dorsal surface of the spinal cord, supplying the outer part of the spinal cord including the major part of the dorsal horns [28].

2.4. Extrinsic arteries

The majority of the intrinsic arteries (central arteries) arise from the ventral spinal artery. In place of fusion of bilateral vertebral arteries, the ventral spinal artery is in cranial connection with both of them or only with one of them. The ventral spinal artery is located in the ventral median fissure and continues throughout the whole length of the spinal cord [28]. To the level of the thoracic part of the spinal cord decreases the diameter of the ventral spinal artery. It is constant in the thoracic part and more caudally is the artery hardly recognizable. At the level of joining of the artery of Adamkiewicz, the ventral spinal artery reaches its smallest diameter. Caudally to this junction, the ventral spinal artery becomes larger in diameter.

The lower thoracic, lumbar and sacral part of the spinal cord are supplied by a spinal branch with larger diameter which is known as the artery of Adamkiewicz (*arteria radicularis magna*, great radicular artery). The typical hairpin curve of the artery of Adamkiewicz is caused by a different speed of growth of the spinal cord and vertebral column [31]. Before accompanying the corresponding nerve root, the artery courses cranially form an oblique angle. After giving off a small cranially directed branch, the artery reaches the ventral median fissure. The artery of Adamkiewicz is continuing vertically, turns caudally in an acute angle and reaches the ventral median fissure. This anatomical arrangement predisposes the direction of the blood flow into the caudal segments of the spinal cord [32].

In the dorsal lateral grooves of the spinal cord, the longitudinal dorsal spinal arteries run. The cranial connection of the dorsal spinal arteries is to the posterior inferior cerebellar arteries or to the vertebral arteries. The diameter of the arteries varies according to the part of the spinal cord.

2.5. Intrinsic arteries

The vascular network formed by the intrinsic arteries consists of the central and peripheral system. The central system supplies the ventral two-thirds of the spinal nervous tissue. The blood flow is centrifugal from the ventral spinal artery. The blood of the central system supplies the base of the dorsal columns of the white matter, the ventral grey matter, dorsal columns of the white matter, ventral part of the dorsal grey matter and inner half of the ventral and lateral columns of the white matter. The peripheral system conveys the blood through the pial arterial plexus and dorsal spinal arteries centripetally. Dorsal columns of the white matter, dorsal part of the dorsal grey matter and outer part of the ventral and lateral columns of the white matter are supplied by the peripheral system [28].

The ratio of blood supply to spinal cord by central and peripheral system is variable according to the part of the spinal cord. The central system is large in the lumbar and cranial sacral segments, the central and peripheral system are large in the cervical segments and the peripheral system is large in the thoracic segments [28].

http://dx.doi.org/10.5772/intechopen.69119

3. Human

To the most important factors affecting the severity of degree and clinical results of the spinal cord injury belongs the spinal cord ischemia. The planning of surgical and endovascular interventions cannot be successful with the detailed knowledge of the arterial system of the spinal cord [28].

3.1. Cervical part of the spinal cord

The cervical spinal cord is supplied by segmental spinal branches arising from the vertebral, deep cervical and ascending cervical arteries [33]. The cranial three to five spinal branches in the cervical part originate from the vertebral arteries and in the caudal part from the deep cervical, ascending cervical arteries and the supreme intercostal artery [34]. Occasionally some caudal spinal branches in the cervical part may arise as direct branches from the subclavian artery [32]. The branches of the ascending cervical artery fuse together with the branches of the vertebral and deep cervical artery before giving of the spinal branches [35].

In the literature, high variability in the origin of the vertebral artery was described [36]. The vertebral arteries originate from the first parts of the subclavian arteries. After they leave the thoracic cavity through the thoracic inlet, they enter the transverse canal at the level of the sixth cervical vertebra and continue inside the canal to the head. They enter the vertebral canal via the transverse foramen of the atlas.

The arterial blood to the cervical spinal cord convey eight to ten unpaired ventral spinal branches originating from dorsal segmental branches [32]. One of them is located at the level of the sixth cervical vertebra, and one to three cranially and caudally to this spinal branch [37]. They enter the ventral spinal artery. The largest of them is located at the cervical enlargement at the level of the sixth or fifth cervical segment (rarely at the fourth or seventh cervical segment) [27, 34] and it is arising from the deep cervical artery (in case of rare location from the ascending cervical artery or costocervical trunk). The second largest spinal branch is positioned at the level of the third cervical segment (rarely at the second or fourth cervical segment) [34].

The ventral spinal artery is cranially connected to the both vertebral arteries [38, 39]. It has a form of an uninterrupted trunk located in the ventral median sulcus, will often bifurcate and vary in size [40].

Two plexiform dorsal spinal arteries are cranially connected to the vertebral arteries or posterior inferior cerebellar arteries [41–43]. Their diameter is the largest in the cervical region [32].

3.2. Thoracolumbar part of the spinal cord

The thoracic part of the spinal cord is supplied by spinal branches originating from the intercostal arteries [33]. From the dorsal surface of the thoracic aorta, nine pairs of segmental intercostal arteries and one pair of subcostal arteries arise [34]. Two to four small spinal branches enter the ventral spinal artery in the thoracic part [32].

The segmental spinal branches supplying the lumbar part of the spinal cord are branches of the paired lumbar arteries, iliolumbararteries, and the lowest lumbar artery. In the sacral part convey the arterial blood the lateral sacral arteries [33]. The lumbar arteries are present in four pairs with origin from the dorsal surface of the abdominal aorta. The iliolumbar artery branches either from the abdominal aorta and median sacral artery [34].

The arterial system in the middle segments of the thoracic part is poorly developed. In the majority of cases, it is supplied only by one spinal branch originating from the seventh intercostal artery. The ventral spinal artery is discontinuous in this region [44].

The second spinal branch participating on the blood supply of the thoracic and lumbar part is the artery of Adamkiewicz [45]. It is the largest spinal branch with variable level of its origin [46–48]. In 75% of cases, its origin is located at the level between ninth and twelfth thoracic segment, in 15% of cases between the fifth and eighth thoracic segment and in 10% of cases between the first and second lumbar segment [42, 47]. As left-sided artery, it was described in 80% of cases [49]. The artery of Adamkiewicz supplies 68% of perfusion to the caudal thoracic and cranial lumbar part of the spinal cord [50]. This artery forms a typical hairpin curve and thereafter reaches the ventral spinal artery [47, 51]. In the case of the origin of the artery of Adamkiewicz at the level between the fifth and eighth thoracic segment, an accessory artery known as the Desproges-Gotteron artery (cone artery) can be present. Its ordinary position is at the level between the second and fifth lumbar segment [52]. Most frequent level is the second lumbar segment, but it may vary between the eighth thoracic and fourth lumbar segment [42].

The spinal branches in the sacral part of the spinal cord originate from the lateral sacral arteries and the median sacral artery. They supply the nerve roots and sacral nerves. The extensive anastomosis on the ventral surface of the sacrum is formed by spinal branch directed to the first ventral sacral foramen and spinal branches, which enter the second, third and fourth ventral sacral foramina [34].

Along the whole length of the spinal cord, the ventral spinal artery receives six to eight significant ventral branches. The ventral branches have a larger diameter and are less numerous as the dorsal branches [32, 33]. The bifurcation is formed when the ventral branches enter the ventral spinal artery from both sides at the same segment. Such bifurcation forms a diamond-shaped pattern. The diameter of the ventral spinal artery correlates to the relative amount of the grey matter [34]. Its widest point is at the junction with the artery of Adamkiewicz and the second widest at the cervical enlargement. The ventral spinal artery is narrowest in the middle segments of the thoracic part of the spinal cord [32].

Two dorsal spinal arteries are located on the dorsal surface of the spinal cord and they receive eight to sixteen significant dorsal branches. They join at the caudal level of the *conus medullaris* with the ventral spinal artery to form the conus basket which extends to the *filum terminale* [53]. The ventral spinal artery gives off two dorsally directed terminal branches which anastomose with each dorsal spinal artery at the level of the fifth sacral segment [54]. The diameter of the dorsal spinal arteries is smaller in the thoracic region than in lumbar region [32].

4. Experimental animals

4.1. Pig

Despite the widespread and intensive use of the pig as an animal model in the experimental spinal cord injury, there is a gap in the information of its spinal cord arterial supply [55, 56].

4.1.1. Cervical part of the spinal cord

The vertebral arteries arise uniformly as the first branches on both sides from the costocervical trunk [57] or variably [26]. The right vertebral artery is the second branch arising from the right costocervical trunk and the left vertebral artery is the third branch with the origin from the left subclavian artery [26].

The spinal branches supplying the cervical part of the spinal cord are arising segmentally from the bilateral vertebral arteries [57]. The ventral spinal artery is cranially connected by means of two ventral branches to the vertebral arteries in the place of formation of the basilar artery [57, 58]. The ventral spinal artery is uninterrupted and entered on average by six ventral branches [58].

4.1.2. Thoracolumbar part of the spinal cord

In the literature, the variable arrangement of the dorsal intercostal arteries is described. The first dorsal intercostal artery has origin from the vertebral artery [57], supreme intercostal artery [59] or it is not present [26]. The second dorsal intercostal artery is absent in generally [26] or it arises in some cases from the dorsal scapular artery or regularly from the vertebral artery [59]. The third, fourth and the fifth dorsal intercostal arteries are arising from the supreme intercostal artery [26] or from the dorsal surface of the thoracic aorta as independent branches [59]. The rest of the dorsal intercostal arteries of the same level originate from the dorsal surface of the thoracic aorta as individual branches [55, 56, 60] or they form a common trunk of origin [59].

The arrangement of the lumbar arteries is more constant. They are present as six [55, 56, 60] or five pairs [26, 59] with independent origin from the dorsal surface of the abdominal aorta. In the case of presence of five pairs, the sixth pair of the lumbar arteries branches from the median sacral artery [26].

In the thoracic part, the ventral spinal artery receives on average eight ventral branches and in the lumbar part five ventral branches [58]. In the majority of cases, the artery of Adamkiewicz originates from the caudal part of the abdominal aorta [61]. The range of origin of the artery of Adamkiewicz is from the first lumbar to the first sacral segment. In 87% of cases, the origin is from the third to the fifth lumbar artery [62].

In the pig, the arterial blood supply to the thoracolumbar part of the spinal cord is enriched by the collateral blood flow, which consists of two parts. The first part is coming from the internal thoracic and subscapular artery through the wall of the chest and abdomen. The second part presents the blood coming from the median sacral artery through its spinal branches to the most caudal part of the spinal cord [55].

4.2. Dog

In studies of several spinal cords, diseases serve as an animal model the dog. In the research area, a lot of papers dealing with experimental ischemia in the dog have been published [29].

4.2.1. Cervical part of the spinal cord

The origin of the vertebral artery is located on the subclavian artery before the origin of the costocervical trunk on both sides [26, 63]. The collateral blood flow is formed by connections between four to five muscular branches of the costocervical trunk and each vertebral artery and between the bilateral superficial cervical artery and vertebral artery [64].

The spinal branches originating segmentally from the vertebral artery at each of the seven intervertebral foramina supply the cervical spinal cord. To the largest spinal branches belongs the first [65] and the third [63, 65], and in some cases also the fourth branch [66]. From the dorsal scapular artery that branches from the costocervical trunk originates the eighth spinal branch [66].

The unpaired ventral spinal artery extending the whole length of the spinal cord is formed by the fusion of the ventral branches of the spinal branches [65, 66]. In 67% of cases, the ventral spinal artery is in some segments of the cervical spinal cord doubled [65, 67]. The ventral spinal artery is high variable in its cranial connection: in 20% of cases, it is the point of fusion of bilateral vertebral arteries; in 20% of cases, it is from the bilateral vertebral arteries in the place of formation of the basilar artery; in 40% of cases, it is from the right vertebral artery and in the last 20% of cases, it is from the left vertebral artery [29]. The cranial connection of the ventral spinal artery in the area of formation of basilar artery is also described as constant [63, 68]. The ventral spinal artery is joining 88% of all the possible ventral branches. The diameter of the ventral spinal artery is larger in the cervical than in the thoracic part [65].

From the spinal branches arising dorsal branches form on the dorsolateral surface of the spinal cord an irregular arrangement [66] or continuous dorsal spinal arteries [65] throughout the entire spinal cord. On the dorsal surface of the spinal cord, four dorsal spinal arteries are present. The larger lateral dorsal spinal arteries are located in the dorsal lateral grooves and the smaller medial dorsal spinal arteries in the dorsal intermediate grooves. The dorsal spinal arteries are cranially connected to the basilar artery (60% of cases) or to the rostral cerebellar artery (40% of cases) [29]. The dorsal spinal arteries are formed throughout the whole length of the spinal cord and they receive 88.1% of all the possible dorsal branches. The dorsal branches in the cervical part are originating from the ventral branches [65].

The several anastomoses between the dorsal and ventral spinal arteries form the spinal arterial ring which is higher density in cervical than in the thoracic part of the spinal cord [29].

4.2.2. *Thoracolumbar part of the spinal cord*

The dorsal intercostal arteries are present in a number of twelve pairs [66]. The first dorsal intercostal artery branches from the costocervical trunk [26, 66]. The second, third [26, 66] and in some cases the fourth dorsal intercostal artery are coming from the thoracic vertebral artery, which originates from the costocervical trunk [66]. One study described the second, occasionally the third and the fourth dorsal intercostal artery as branches of a common trunk with the fifth right dorsal intercostal artery [69]. The rest of dorsal intercostal arteries originate from the dorsal surface of the thoracic aorta, by which the left-sided arise more cranially than the right-sided [66].

The total number of lumbar arteries is seven pairs. The first and second pair of lumbar arteries are arising from the thoracic aorta because of the attachment of the diaphragm to the third and fourth lumbar vertebrae. The next five pairs are originating from the dorsal surface of the abdominal aorta whereby the left-sided arise cranially to the right-sided. More caudally, the origins of the arteries at the same level become closer and the last pairs of the lumbar arteries originate by means of a common trunk. The last seventh pair arises from a common trunk with the median sacral artery from the terminal part of the abdominal aorta [26, 66]. The median sacral artery represents the caudal continuation of the abdominal aorta. From the median sacral artery originating sacral branches give off spinal branches which enter the ventral spinal artery [66].

The spinal branches with corresponding nerve root enter the vertebral canal on the right side in 72% of cases and on the left side in 62% of cases. The diameter of the ventral spinal artery becomes smaller caudally to the lower cervical part of the spinal cord [29, 65] and it is smaller in the thoracic part in comparison with the lumbar part of the spinal cord. In the thoracic part enter the ventral spinal artery 31.2% from all the possible ventral branches and in the lumbar part 45% of them [65].

In 50% of cases, the artery of Adamkiewicz arises from the left fifth lumbar artery and in 50% of cases, it is absent. It represents the arterial supply to the ventral two-thirds of the caudal half of the spinal cord. In the place of fusion with the ventral spinal artery, it gives off a cranially directed branch to the thinned ventral spinal artery [29]. The arrangement, the side and the level of origin of the artery of Adamkiewicz was observed as high variable [65].

The dorsal spinal arteries in the thoracic part receive 49.6% all the possible dorsal branches and in the lumbar part 60% of them. The diameter of the dorsal spinal arteries is smaller in comparison with the cervical and lumbar part [65].

In the lumbar part, the density of the spinal arterial ring is higher than in the thoracic part [29, 65].

4.3. Cat

Despite the cat does not belong to the most frequently used experimental model in the study of the spinal cord injuries, it has an important place as an animal model in such studies.

4.3.1. Cervical part of the spinal cord

The vertebral artery is bilaterally the first branch arising from the subclavian artery [26, 57, 70].

The ventral spinal artery is cranially connected to the both vertebral arteries in place of their fusion and runs throughout the entire length of the spinal cord [68, 70, 71]. On some distance, it is positioned deeper in the ventral median fissure [70]. 80.6% of all possible ventral branches enter the ventral spinal artery. The first ventral branch has the largest diameter [65].

Two longitudinal dorsal spinal arteries run on the dorsal surface [65, 70]. 95.6% of all possible dorsal branches enter the dorsal spinal arteries [65].

4.3.2. Thoracolumbar part of the spinal cord

The total number of dorsal intercostal arteries is twelve pairs. The first dorsal intercostal artery originates from the costocervical trunk, and the second and third arise from the supreme intercostal artery. The remaining pairs originate independently from the dorsal surface of the thoracic aorta [26].

There are present six pairs of lumbar arteries with origin from the dorsal surface of the abdominal aorta [26, 57, 71]. The median sacral artery gives off the last seventh pair [26, 71].

The diameter of the ventral spinal artery in the thoracic part is smaller in comparison with the lumbar and cervical part [65, 70]. From all possible ventral branches enter the ventral spinal artery in the thoracic part 55.4% and in the lumbar part 42.8% of them [65]. The frequency of the left-sided ventral branches is doubled in comparison with the frequency of the right-sided [71].

Till now, it was described the different anatomical arrangement of the artery of Adamkiewicz in the literature. The origin from the left fourth lumbar artery is designated as uniform [70]. Another one source describes in 80% of cases only a larger ventral branch connecting the ventral spinal artery. In 80% of cases, it arises from the fourth lumbar artery and in 20% of cases from the third lumbar artery. As left-sided artery, it is present in 70% of cases [65]. The similar arrangement describes other work. In 70% of cases, a larger ventral branch originates from the fourth lumbar artery and in 30% of cases from the third lumbar artery. An accessory large ventral branch is formed in 50% of cases [71].

The dorsal spinal arteries in the thoracic part are smaller in diameter than in the cervical and lumbar part. 56.9% of possible dorsal branches enter the dorsal spinal arteries in the thoracic part and 70.7% in the lumbar part [65]. Bradshaw [70] described four dorsal spinal arteries located caudally to the level from the fifth to the seventh cervical vertebra. One pair is located medially and one pair laterally. In the lumbar part, also two dorsolaterally positioned dorsal spinal arteries with no side predominance are observed [71].

The spinal arterial ring is formed only in one-half of cases [71]. The density of the spinal arterial ring in the thoracic part is smaller in comparison with the cervical and lumbar part of the spinal cord [65, 72].

4.4. Rabbit

The rabbit is one of the most used species of experimental animals serving for the study the pathophysiology of the spinal cord diseases and neuroprotective drugs effect on damaged nervous tissue [73].

4.4.1. Cervical part of the spinal cord

Till now, not only constant origin of the vertebral artery from the subclavian artery [74, 75] is described but also variations. In 86% of cases, the left-sided vertebral artery arises from the left subclavian artery, in 10% of cases from the aortic arch and in 4% of cases from a common trunk with the left descending scapular artery with the origin from the aortic arch. In 98% of cases, the right-sided vertebral artery originates from the right subclavian artery. In the remaining 2% of cases, it is formed by the fusion of two branches. The first branch originates from the right subclavian artery and the second branch from the common trunk with the right superficial cervical artery from the common carotid artery [76]. The fusion of both vertebral arteries is in 50% of cases without a gap, in 30% of cases with one longitudinal gap and in 20% of cases with two oval gaps [77].

The unpaired and continuous ventral spinal artery is connected in the area of both vertebral arteries fusion to the right vertebral artery in 40% of cases, to the left vertebral artery in 35% of cases or to the bilateral vertebral arteries in 25% of cases [77]. Another study describes the cranial connection of the ventral spinal artery to the bilateral vertebral arteries in all studied specimens [68]. The left-sided ventral branches enter the ventral spinal artery in 53.8% of cases and the right-sided in 46.2% of cases [77]. The doubled ventral spinal artery without recording its cranial connection is described too [67].

On the dorsal surface of the spinal cord are located two or none dorsal spinal arteries, but the frequency of the presence or absence is not noted [77]. Cranially they are connected to the corresponding vertebral artery, or posterior inferior cerebellar artery [67] or none [77]. The dorsal branches entering the dorsal spinal arteries are present in the same frequency as the ventral branches which are joining the ventral spinal artery [77].

4.4.2. Thoracolumbar part of the spinal cord

The thoracic part of the spinal cord receives the arterial blood from the spinal branches originating from 12 pairs of the dorsal intercostal arteries and from the costoabdominal artery. The dorsal intercostal arteries with the origin from the dorsal surface of the thoracic aorta are present as nine pairs in 70% of cases, as eight pairs in 20% of cases and as 10 pairs in 10% of cases. The rest of them is coming from the supreme intercostal artery. The arrangement of the dorsal intercostal arteries of the same level is high variable [78]. The dorsal intercostal arteries originate as paired segmental branches arising from the dorsal surface of the thoracic aorta [74, 75, 79].

The lumbar arteries originating from the dorsal surface of the abdominal aorta which are supplying the lumbar part of the spinal cord are present in a number of six pairs in 90% of cases and in a number of five pairs in 10% of cases [78]. The last seventh pair branches from the median sacral artery [78, 79]. The lumbar arteries of the same level arise as independent branches in 60% of cases and in 40% of cases is their origin high variable [78]. Also, seven pairs of segmental lumbar arteries with origin from the abdominal aorta in all the studied specimens are possible [74, 75].

The continuous and unpaired ventral spinal artery is positioned in the ventral median fissure [78]. In the lumbar part of the spinal cord is possible the presence of three parallel longitudinal ventral spinal arteries [68]. In the thoracic part of the spinal cord, the left-sided ventral branches joining the ventral spinal artery are present in 71% of cases and the right-sided in 29% of cases. In the lumbar part in 62.5% of cases are formed the left-sided ventral branches and in 37.5% of cases the right-sided [78].

The artery of Adamkiewicz originates in 50% of cases from the right-sided and in 50% of cases from the left-sided sixth lumbar artery [78]. In the other work, the presence and the level of the origin of the artery of Adamkiewicz is not recorded [68].

In 70% of cases, two dorsal spinal arteries are present, in 20% of cases no one and in 10% of cases three. The left-sided dorsal branches joining the dorsal spinal arteries are present in the thoracic part in 60.5% of cases and right-sided in 39.5% of cases, in the lumbar part the left-sided in 52.5% of cases and right-sided in 47.5% of cases [80].

4.5. Rat

The rat is matchless the most frequently used experimental animal in the study of spinal cord injury and absolutely in the study of the spinal cord ischemia.

4.5.1. Cervical part of the spinal cord

The both vertebral arteries arise from the subclavian arteries as the second branches after the origin of the costocervical trunks [81, 82]. The cranial connection of the ventral spinal artery is by means of two ventral branches to the bilateral vertebral arteries in the place of the formation of the basilar artery [81, 83].

Two ventral spinal arteries [67, 82] are in connection with the corresponding vertebral artery [82] or without cranial connection [67]. Three to four ventral branches join the ventral spinal artery in the cervical part of the spinal cord [68, 83]. The last cervical segments have rich arterial blood supply by means of the ventral branches [84]. On each side originates from the vertebral artery the corresponding dorsal spinal artery [81].

4.5.2. Thoracolumbar part of the spinal cord

The dorsal intercostal arteries are present in a number of twelve pairs. The first, second and third pair arise from the supreme intercostal artery which originates from the costocervical trunk. From the dorsal surface of the thoracic aorta, nine pairs of the dorsal intercostal arteries arise independently. The dorsal costoabdominal artery represents the last segmental pair [82]. The origin of the dorsal intercostal arteries at the same level from the dorsal surface of the thoracic aorta was described also by means of a common trunk [85].

Five pairs of lumbar arteries with independent origin [82] or with origin by means of a common trunk [85] arise from the dorsal surface of the abdominal aorta.

The cranial thoracic part is supplied by means of a poor arterial system formed by segmental branches in this area [84]. The unpaired and uninterrupted ventral spinal artery runs subdurally in the ventral median fissure from the level of the tenth thoracic to the fourth lumbar vertebra [86]. Two to three [84] or three to four ventral branches enter the ventral spinal artery. The majority of them are located at the level from the tenth thoracic to the second lumbar vertebra [81]. Three ventral branches are present in 58% of cases, four ventral branches in 27% of cases and five ventral branches in 15% of cases. For the ventral branches, the left-sided predominance is typical [86]. The ventral spinal artery continues to the beginning of *filumterminale* [83, 84].

The artery of Adamkiewicz is relatively constant in topography with right-sided predominance and it is accompanied in the cranial region of the lumbar part of the spinal cord with one or two small branches [84]. The caudal thoracic, lumbar and sacral part of the spinal cord [68, 72, 83, 86] are supplied at the level of the second or third lumbar vertebra [83, 86] by a large segmental artery (the artery of Adamkiewicz).

On the dorsal surface of the spinal cord, two irregular dorsal spinal arteries with irregular loops between each other are formed [83]. On the dorsal surface of the spinal cord, except the cervical part, three additional dorsal spinal arteries are located: two lateral dorsal spinal arteries and one median dorsal spinal artery. The additional arteries are interconnected by means of the transverse anastomotic circle [81]. The dorsal branches are higher in number and equally distributed, but of a smaller diameter than the ventral branches. They are more frequently present at the cervical and lumbar enlargements [84].

In the most caudal segments, the arterial plexuses are present on the surface of the spinal cord [84]. The arterial anastomoses are present in half of cases and the arterial network is of lesser density in thoracic part than in the cervical and lumbar part [72].

5. Conclusions

Detailed knowledge of the anatomy of the arteries contributing to the spinal cord blood supply plays an important role in the management of treatments of several diseases of the thoracic and/or thoracolumbar aorta, which may impact the spinal cord blood supply [87]. The possible present collateral arterial system is of great importance as a compensatory mechanism of the spinal cord blood supply in the cases of large arteries occlusion. The supply from one source can decrease when another one is increased or the other way [88]. A high risk of the spinal cord ischemia can be caused by the occlusion of the segmental arteries and rupture of a possible collateral system of the spinal cord blood supply [89].

In the pig, each dorsal intercostal and lumbar artery conveys the blood into the ventral spinal artery. In less than quarter of studied pigs was monitored, the manifested paraplegia after the ligation of the descending aorta [90] or segmental arteries was known as critical. The arteries intended for another spinal cord segments replace the arterial blood flow to the segments with interrupted arterial supply. All dorsal intercostal and lumbar arteries participate on the arterial blood supply of the spinal cord and on the formation of the collateral blood flow [91]. The differences in arterial pattern concerning the collateral blood circulation must also be considered by the interpretation of the results of the experimental studies [59]. Several arteries with the variable place of their origin supply the spinal cord in the pig. The median sacral artery represents an important source of the pig spinal cord arterial supply which differentiates it from the human spinal cord blood supply.

The use of the dog in the study of experimental spinal cord ischemia predetermines: the blood supply of the ventral two-thirds of the spinal cord in the caudal cervical part by main arteries with a greater diameter and the right-sided predominance of the segmental arteries [29, 92]. In comparison with the rabbit, the spinal cord blood supply in the dog is more similar to the humans. The clamping of the thoracic aorta is an adequate method to induce a spinal cord ischemia of respective spinal cord segments [29, 90].

In cat, the abdominal aorta gives rise to one or more branches with larger calibre, which convey the arterial blood to the lumbar and sacral part of the spinal cord [93]. This arterial pattern resembles that in human.

The effect of numerous neuroprotective drugs and postoperative outcomes is studied in rabbit, which serves as a better experimental model for such experiments. The homosegmental blood supply from the abdominal aorta caudally to the origin of renal arteries with minimal or no intraspinal collateral arterial system and that the neurological outcomes do not differ significantly from the neurological signs in other species of experimental animals predetermine the use of the rabbit in neurological research [93, 94]. To induce the spinal cord ischemia in rabbit, it is necessary to ligate only the abdominal aorta [90].

In rats, the heterosegmental blood supply is formed in the lower thoracic, lumbar and sacral spinal cord. To obtain the spinal cord ischemia in rat, it is necessary the ligation of the descending aorta [90].

Before the appropriate animal model is chosen, several aspects must be considered. It is very difficult to find an ideal animal model because of different occurrence of variable advantages and disadvantages [93]. Each aspect and step before starting an experimental model must be considered. It is recommended to perform a pilot study which helps the scientist to determinate what is awaiting from each animal model and identifies the optimal way for reduction of the animal use and necessary experimental time [95].

Author details

Slavka Flesarova and David Mazensky*

*Address all correspondence to: mazenskyd@azet.sk

Department of Anatomy, Histology and Physiology, University of Veterinary Medicine and Pharmacy in Kosice, Kosice, The Slovak Republic

References

[1] Bosmia AN, Hogan E, Loukas M, Tubbs RS, Cohen-Gadol AA. Blood supply to the human spinal cord: Part I. Anatomy and hemodynamics. Clinical Anatomy. 2015;**28**:52-64. DOI: 10.1002/ca.22281

[2] Pickelsimer E, Shiroma EJ, Wilson DA. Statewide investigation of medically attended adverse health conditions of persons with spinal cord injury. The Journal of Spinal Cord Medicine. 2010;**33**:221-231. DOI: 10.1080/10790268.2010.11689699

[3] Martirosyan N, Feuerstein JS, Theodore N, Cavalcanti DD, Spetzler RF, Preul MC. Blood supply and vascular reactivity of the spinal cord under normal and pathological conditions. Journal of Neurosurgery. Spine. 2011;**15**:238-251. DOI: 10.3171/2011.4.SPINE10543

[4] Santillan A, Nacarino V, Greenberg E, Riina HA, Gobin YP, Patsalides A. Vascular anatomy of the spinal cord. Journal of Neurointerventional Surgery. 2012;**4**:67-74. DOI: 10.1136/neurintsurg-2011-010018

[5] Lee JH, Jones CF, Okon EB, Anderson L, Tigchelaar S, Kooner P, Godbey T, Chua B, Gray G, Hildebrandt R, Cripton P, Tetzlaff W, Kwon BK. A novel porcine model of traumatic thoracic spinal cord injury. Journal of Neurotrauma. 2013;**30**:142-159. DOI: 10.1089/neu.2012.2386

[6] Jones CF, Lee JH, Burstyn U, Okon EB, Kwon BK, Cripton PA. Cerebrospinal fluid pressures resulting from experimental traumatic spinal cord injuries in a pig model. Journal of Biomechanical Engineering. 2013;**135**:101005. DOI: 10.1115/1.4025100

[7] Šulla I, Bačiak L, Juránek I, Cicholesová T, Boldižár M, Balik V, Lukáčová N. Assessment of motor recovery and MRI correlates in a porcine spinal cord injury model. ActaVeterinaria Brno. 2014;**83**:393-397. DOI: 10.2754/avb201483040393

[8] Fujimaki Y, Kawahara N, Tomita K, Murakami H, Ueda Y. How many ligations of bilateral segmental arteries cause ischemic spinal cord dysfunction? An experimental study using a dog model. Spine—Affiliated Society Meeting Abstracts. 2006;**31**:E781–E789. DOI: 10.1097/01.brs.0000238717.51102.79

[9] Bozynski CC, Vasquez L, O'Brien DP, Johnson GC. Compressive myelopathy associated with ectasia of the vertebral and spinal arteries in a dog. Veterinary Pathology. 2012;**49**:779-783. DOI: 10.1177/0300985811415704

[10] Sarmento CA, Rodrigues MN, Bocabello RZ, Mess AM, Miglino MA. Pilot study: Bone marrow stem cells as a treatment for dogs with chronic spinal cord injury. Regenerative Medicine Research. 2014;**2**:9. DOI: 10.1186/2050-490X-2-9

[11] Tsitsopoulos PD, Tsonidis CA, Tsoleka KD, Tavridis GN. The effect of radicular arteries ligation on the spinal cord blood supply in cats. British Journal of Neurosurgery. 1995;**9**:227-232. DOI: 10.1080/02688699550041610

[12] Murphy MN, Ichiyama RM, Iwamoto GA, Mitchell JH, Smith SA. Exercise pressor reflex function following acute hemi-section of the spinal cord in cats. Frontiers in Physiology. 2013;**4**:3. DOI: 10.3389/fphys.2013.00003

[13] Gu B, Olejar KJ, Reiter JP, Thor KB, Dolber PC. inhibition of bladder activity by 5-hydroxytryptamine1 serotonin receptor agonists in cats with chronic spinal cord injury. Journal of Pharmacology and Experimental Therapeutics. 2004;**310**:1266-1272. DOI: 10.1124/jpet.103.063842

[14] Leonard AV, Thornton E, Vink R. NK1 receptor blockade is ineffective in improving outcome following a balloon compression model of spinal cord injury. PLoS One. **9**:e98364. DOI: 10.1371/journal.pone.0098364

[15] Fang B, Li XM, Sun XJ, Bao NR, Ren XY, Lv HW, Ma H. Ischemic preconditioning protects against spinal cord ischemia-reperfusion injury in rabbits by attenuating blood spinal cord barrier disruption. International Journal of Molecular Sciences. 2013;**14**:10343-10354. DOI: 10.3390/ijms140510343

[16] Mechírová E, Danielisová V, Domoráková I, Danková M, Stebnický M, Mičková H, Burda J. Bradykinin preconditioning affects the number of degenerated neurons and the level of antioxidant enzymes in spinal cord ischemia in rabbits. Acta Histochemica. 2014;**116**:252-257. DOI: 10.1016/j.acthis.2013.07.010

[17] Ouyang H, Galle B, Li J, Nauman E, Shi R. Biomechanics of spinal cord injury: A multimodal investigation using ex vivo guinea pig spinal cord white matter. Journal of Neurotrauma. 2008;**25**:19-29. DOI: 10.1089/neu.2007.0340

[18] Galle B, Ouyang H, Shi R, Nauman E. A transversely isotropic constitutive model of excised guinea pig spinal cord white matter. Journal of Biomechanics. 2010;**43**:2839-2843. DOI: 10.1016/j.jbiomech.2010.06.014

[19] Sun W, Fu Y, Shi Y, Cheng J-X, Cao P, Shi R. Paranodal myelin damage after acute stretch in guinea pig spinal cord. Journal of Neurotrauma. 2012;**29**:611-619. DOI: 10.1089/neu.2011.2086

[20] Chung WH, Lee JH, Chung DJ, Yang WJ, Lee AJ, Choi CB, Chang HS, Kim DH, Chung HJ, Suh HJ, Hwang SH, Han H, Do SH, Kim HY. Improved rat spinal cord injury model using spinal cord compression by percutaneous method. Journal of Veterinary Science. 2013;**14**:329-335. DOI: 10.4142/jvs.2013.14.3.329

[21] Losey P, Anthony DC. Impact of vasculature damage on the outcome of spinal cord injury: A novel collagenase-induced model may give new insights into the mechanisms involved. Neural Regeneration Research. 2014;**9**:1783-1786. DOI: 10.4103/1673-5374.143422

[22] Soubeyrand M, Badner A, Vawda R, Chung YS, Fehlings MG. Very high resolution ultrasound imaging for real-time quantitative visualization of vascular disruption after spinal cord injury. Journal of Neurotrauma. 2014;**31**:1767-1775. DOI: 10.1089/neu.2013.3319

[23] Gaviria M, Haton H, Sandillon F, Privat A. A mouse model of acute ischemic spinal cord injury. Journal of Neurotrauma. 2002;**19**:205-221. DOI: 10.1089/08977150252806965

[24] Awad H, Ankeny DP, Guan Z, Wei P, McTigue DM, Popovich PG. A mouse model of ischemic spinal cord injury with delayed paralysis caused by aortic cross-clamping. Anesthesiology. 2010;**113**:880-891. DOI: 10.1097/ALN.0b013e3181ec61ee

[25] Abematsu M, Tsujimura K, Yamano M, Saito M, Kohno K, Kohyama J, Namihira M, Komiya S, Nakashima K. Neurons derived from transplanted neural stem cells restore disrupted neuronal circuitry in a mouse model of spinal cord injury. The Journal of Clinical Investigation. 2010;**120**:3255-3266. DOI: 10.1172/JCI42957

[26] Nickel R, Schummer A, Seiferle E. The anatomy of the domestic animals, vol. **3**. 1st ed. Berlin: Verlag Paul Parey; 1981. p. 612

[27] Lazorthes G, Gouaze A, Zadeh JO, Santini JJ, Lazorthes Y, Burdin P. Arterial vascularization of the spinal cord. Recent studies of anastomotic substitution pathways. Journal of Neurosurgery. 1971;**35**:253-262. DOI: 10.3171/jns.1971.35.3.0253

[28] Turnbull IM. Chapter 5. Blood supply of the spinal cord: Normal and pathological considerations. Clinical Neurosurgery. 1973;**20**:56-84

[29] Pais D, Casal D, Arantes M, Casimiro M, O'Neill JG. Spinal cord arteries in *Canis familiaris* and their variations: Implications in experimental procedures. Brazilian Journal of Morphological Sciences. 2007;**24**:224-228

[30] Limet R, Lenelle J, Creemers E. Peripheral Vascular Surgery. 1st ed. Ingelheim am Rhein: Boehringer Ingelheim International GmbH; 1994. p. 196

[31] Prince EA, Ahn SH. Basic vascular neuroanatomy of the brain and spine: What the general interventional radiologist needs to know. Seminars in Interventional Radiology. 2013;**30**:234-239. DOI: 10.1055/s-0033-1353475

[32] Crosby E, Gillilian L. Correlative Anatomy of the Nervous System. 1st ed. New York: The Macmillan Company; 1962. p. 731

[33] Yoss RE. Vascular supply of the spinal cord: The production of vascular syndromes. Medical Bulletin (Ann Arbor). 1950;**16**:333-345

[34] Gillilan LA. The arterial blood supply of the human spinal cord. The Journal of Comparative Neurology. 1958;**110**:75-103. DOI: 10.1002/cne.901100104

[35] Bergmann L, Alexander L. Vascular supply of the spinal ganglia. Archives of Neurology and Psychiatry. 1941;**46**:761-782. DOI: 10.1001/archneurpsyc.1941.02280230003001

[36] Lemke AJ, Benndorf G, Liebig T, Felix R. Anomalous origin of the right vertebral artery: Review of the literature and case report of right vertebral artery origin distal to the left subclavian artery. American Journal of Neuroradiology. 1999;**20**:1318-1321

[37] Dunning HS, Wolf HG. The relative vascularity of various parts of the central and peripheral nervous system of the cat and its relation to function. Journal of Comparative Neurology. 1937;**67**:433-450. DOI: 10.1002/cne.900670305

[38] Anderson WD, Kubicek W. The vertebral-basilar system of dog in relation to man and other mammals. The American Journal of Anatomy. 1971;**132**:179-188. DOI: 10.1002/aja.1001320205

[39] Alpagut U, Dayioglu E. Anterior spinal artery syndrome after infrarenal abdominal aortic surgery. The Journal of Cardiovascular Surgery (Torino). 2002;**43**:865-868

[40] Biglioli P, Roberto M, Cannata A, Parolari A, Fumero A, Grillo F, Maggioni M, Coggi G, Spirito R. Upper and lower spinal cord blood supply: The continuity of the anterior spinal artery and the relevance of the lumbar arteries. The Journal of Thoracic and Cardiovascular surgery. 2004;**127**:1188-1192. DOI: 10.1016/j.jtcvs.2003.11.038

[41] Cheshire WP, Santos CC, Massey EW, Howard JF Jr. Spinal cord infarction: Etiology and outcome. Neurology. 1996;**47**:321-330. DOI: 10.1212/WNL.47.2.321

[42] Suh TH, Alexander L. Vascular system of the human spinal cord. Archives of Neurology and Psychiatry. 1939;**41**:659-677. DOI: 10.1001/archneurpsyc.1939.02270160009001

[43] Thron AK, Armin K. Vascular system of the Spinal Cord. Neuroradiological Investigations and Clinical Syndromes. 1st ed. New-York: Springer Verlag; 1988. p. 114. DOI: 10.1007/978-3-7091-6947-6

[44] Bradley WG, Daroff RB, Fenichel GM, Jankovic J. Neurology in Clinical Practice: The Neurological Disorders. 4th ed. Philadelphia: Elsevier. 2004. 2545 p. DOI: 10.1212/WNL.62.9.1657-a

[45] Alleyne CH Jr, Cawley CM, Shengelaia GG, Barrow DL. Microsurgical anatomy of the artery of Adamkiewicz and its segmental artery. Journal of Neurosurgery. 1998;**89:**791-795. DOI: 10.3171/jns.1998.89.5.0791

[46] Uotani K, Yamada N, Kono AK, Taniguchi T, Sugimoto K, Fujii M, Kitagawa A, Okita Y, Naito H, Sugimura K. Preoperative visualization of the artery of Adamkiewicz by intra-arterial CT angiography. American Journal of Neuroradiology. 2008;**29**:314-318. DOI: 10.3174/ajnr.A0812

[47] Domoto S, Kimura F, Asakura T, Nakazawa K, Koike H, Niinami H. Intraspinal collateral circulation to the artery of Adamkiewicz detected with intra-arterial injected computed tomographic angiography. Journal of Vascular Surgery. 2016;**63**:1631-1634. DOI: 10.1016/j.jvs.2015.07.070

[48] Doberstein CA, Bouley A, Silver B, Morrison JF, Jayaraman MV. Ruptured aneurysms of the intradural artery of adamkiewicz: Angiographic features and treatment options. Clinical Neurology and Neurosurgery. 2016;**146**:152-155. DOI: 10.1016/j.clineuro.2016.05.013

[49] Brockstein B, Johns L, Gewertz BL. Blood supply to the spinal cord: Anatomic and physiologic correlations. Annals of Vascular Surgery. 1994;**8**:394-399. DOI: 10.1007/BF02133005

[50] Fanous AA, Lipinski LJ, Krishna C, Roger EP, Siddiqui AH, Levy EI, Leonardo J, Pollina J. The Impact of Preoperative Angiographic Identification of the Artery of Adamkiewicz on Surgical Decision Making in Patients Undergoing Thoracolumbar Corpectomy. Spine (Phila Pa 1976). 2015;**40**:1194-1199. DOI: 10.1097/BRS.0000000000000909

[51] Hyodoh H, Shirase R, Akiba H, Tamakawa M, Hyodoh K, Yama N, Shonai T, Hareyama M. Double-subtraction maximum intensity projection MR angiography for detecting the artery of Adamkiewicz and differentiating it from the drainage vein. Journal of Magnetic Resonance Imaging. 2007;**26**:359-365. DOI: 10.1002/jmri.21024

[52] Novy J, Carruzzo A, Maeder P, Bogousslavsky J. Spinal cord ischemia: Clinical and imaging patterns, pathogenesis, and outcomes in 27 patients. Archives of Neurology. 2006;**63**:1113-1120. DOI: 10.1001/archneur.63.8.1113

[53] Anderson N, Willoughby E. Infarction of the conus medullaris. Annals of Neurology. 1987;**21**:470-474. DOI: 10.1002/ana.410210510

[54] Bolton B. The blood supply of the human spinal cord. Journal of Neurology and Psychiatry. 1939;**2**:137-148

[55] Strauch JT, Spielvogel D, Lauten A, Zhang N, Shiang H, Weisz D, Bodian CA, Griepp RB. Importance of extrasegmental vessels for spinal cord blood supply in a chronic porcine model. European Journal of Cardio-thoracic Surgery. 2003;**24**:817-824. DOI: 10.1016/S1010-7940(03)00460-3

[56] Strauch JT, Lauten A, Zhang N, Wahlers T, Griepp RB. Anatomy of spinal cord blood supply in the pig. The Annals of Thoracic Surgery. 2007;**83**:2130-2134. DOI:10.1016/j.athoracsur.2007.01.060

[57] Popesko P. Anatomy of domesticated animals. 1st ed. Bratislava: Priroda; 1993. p.683

[58] Murray MJ, Bower TC, Carmichael SW. Anatomy of the anterior spinal artery in pigs. Clinical Anatomy. 1992;**5**:452-457. DOI: 10.1002/ca.980050605

[59] Christiansson L, Ulus AT, Hellberg A, Bergqvist D, Wiklund L, Karacagil S. Aspects of the spinal cord circulation as assessed by intrathecal oxygen tension monitoring during various arterial interruptions in the pig. Journal of Thoracic and Cardiovascular Surgery. 2001;**121**:762-772. DOI:10.1067/mtc.2001.112466

[60] de Haan P, Kalkman CJ, de Mol BA, Ubags LH, Veldman DJ, Jacobs MJ. Efficacy of transcranial motor-evoked myogenic potentials to detect spinal cord ischemia during operations for thoracoabdominal aneurysms. The Journal of Thoracic and Cardiovascular Surgery. 1997;**113**:87-101. DOI: 10.1016/S0022-5223(97)70403-3

[61] Wadouh F, Lindeman EM, Arndt CF, Hetzer R, Borst HG. The arteria radicularis magna as a decisive factor influencing spinal cord damage during aortic occlusion. Journal of Thoracic and Cardiovascular Surgery. 1984;**88**:1-10

[62] WissdorfH.DieGefässversorgungderWirbelsäuleunddesRückenmarkesvomHausschwein: (Sus Scrofa F. Domestica L., 1758) [thesis]. Hannover: TierärztlicheHochschule; 1970

[63] Kurtul I, Dursun N, Ozcan S. Pattern of the vertebral artery and its anastomoses in German shepherd dog. KafkasÜniversitesiVeterinerFakültesiDergisi. 2002;**8**:103-106

[64] Whisnant JP, Millikan CH, Wakim KG, Sayre GP. Collateral circulation of the brain of the dog following bilateral ligation of the carotid and vertebral arteries. American Journal of Physiology. 1956;**186**:275-277

[65] Caulkins SE, Purinton PT, Oliver JE Jr. Arterial supply to the spinal cord of dogs and cats. American Journal of Veterinary Research. 1989;**50**:425-430

[66] Evans HE de Lahunta A, de Lahunta A. Miller's Anatomy of the Dog. 4th ed. Missouri: Saunders/Elsevier; 2013. p. 872

[67] Chakravorty BG. Arterial supply of the cervical spinal cord and its relation to the cervical myelopathy in spondylosis. Annals of the Royal College of Surgeons of England. 1969;**45**:232-251

[68] Soutoul JH, Gouaz'e A, Castaing J. The spinal cord arteries of experimental animals. 3. Comparative study of the rat, guinea-pig, rabbit, cat, dog, orang-outang, chimpanzee, with man and fetus. Pathologie Biologie. 1964;**12**:950-962

[69] Reichert FL. An experimental study of the anastomotic circulation in the dog. Bulletin of the Johns Hopkins Hospital. 1924;**35**:385-390

[70] Bradshaw P. Arteries of the spinal cord in the cat. Journal of Neurology, Neurosurgery, and Psychiatry. 1958;**21**:284-289

[71] Chambers G, Eldred E, Eggett C. Anatomical observations on the arterial supply to the lumbosacral spinal cord of the cat. Anatomical Record. 1972;**174**:421-433. DOI: 10.1002/ar.1091740403

[72] Brightman MW. Comparative anatomy of spinal cord vasculature. Anatomical Record. 1956;**124**:264. DOI: 10.1002/ar.1091240209

[73] Lang-Lazdunski L, Matsushita K, Hirt L, WaeberCh, Vonsattel J-PG, Moskowitz MA. Spinal cord ischemia. Development of a model in the mouse. Stroke. 2000;**31**:208-213. DOI: 10.1161/01.STR.31.1.208

[74] Nejedly K. Biology and systematic anatomy of laboratory animals. 1st ed. Prague: SPN; 1965. p. 629

[75] Popesko P, Rajtova V, Horak J. A Colour Atlas of Anatomy of Small Laboratory Animals 1. 1st ed. New York: Saunders; 2003. p. 256

[76] Mazensky D, Danko J. The importance of the origin of vertebral arteries in cerebral ischemia in the rabbit. Anatomical Science International. 2010;**85**:102-104. DOI: 10.1007/s12565-009-0064-8

[77] Mazensky D, Danko J, Petrovova E, Luptáková L, Radoňak J, Schusterová I. Arterial arrangement of the cervical spinal cord in rabbit. Anatomical Science International. 2012;**87**:155-159. DOI: 10.1007/s12565-012-0140-3

[78] Mazensky D, Danko J, Petrovova E, Radoňak J, Frankovičova M. Anatomical study of blood supply to the spinal cord in the rabbit. Spinal Cord. 2011;**49**:525-528. DOI: 10.1038/sc.2010.161

[79] Craige EH. Bensley's practical anatomy of the rabbit. 8th ed. Philadelphia: The Blakistoncompany; 1948. p. 416 DOI: 10.5962/bhl.title.7305

[80] Mazensky D, Danko J, Petrovova E, Mechirova E, Prokes M. Arterial peculiarities of the thoracolumbar spinal cord in rabbit. AnatomiaHistologia Embryologia. 2013;**43**:346-351. DOI: 10.1111/ahe.12081

[81] Paxinos G, Watson C. The Rat Brain in Stereotaxic Coordinates. 7th ed. San Diego: Academic Press; 2013. p. 472

[82] Popesko P, Rajtova V, Horak J. A Colour Atlas of Anatomy of Small Laboratory Animals 2. 1st ed. New York: Saunders; 2003. p. 253

[83] Woollam DHM, Millen JW. The arterial supply of the spinal cord and its significance. Journal of Neurology, Neurosurgery and Psychiatry. 1955;**18**:97-102. DOI: 10.1136/jnnp.18.2.97

[84] Tveten L. Spinal cord vascularity IV. The spinal cord arteries in the rat. Acta Radiologica: Diagnosis. 1976;**17**:385-398. DOI: 10.1177/028418517601700401

[85] Hebel R, Stromberg MW. Anatomy of the laboratory rat. New York: Williams and Wilkins; 1976. p. 173

[86] Schievink WI, Luyendijk W, Los JA. Does the artery of Adamkiewicz exist in the albino rat? Journal of Anatomy. 1988;**161**:95-101

[87] Melissano G, Civilini E, Bertoglio L, Calliari F, Campos Moraes Amato A, Chiesa R. Angio-CT imaging of the spinal cord vascularisation. European Journal of Vascular and Endovascular Surgery. 2009;**39**:436-440. DOI: 10.1016/j.ejvs.2009.11.026

[88] Griepp EB, Di Luozzo G, Schray D, Stefanovic A, Geisbüsch S, Griepp RB. The anatomy of the spinal cord collateral circulation. Annals of Cardiothoracic Surgery. 2012;**1**:350-357. DOI: 10.3978/j.issn.2225-319X.2012.09.03

[89] Czerny M, Eggebrecht H, Sodeck G, Verzini F, Cao P, Maritati G, Riambau V, Beyersdorf F, Rylski B, Funovics M, Loewe C, Schmidli J, Tozzi P, Weigang E, Kuratani T, Livi U, Esposito G, Trimarchi S, van den Berg JC, Fu W, Chiesa R, Melissano G, Bertoglio L, Lonn L, Schuster I, Grimm M. Mechanisms of symptomatic spinal cord ischemia after TEVAR: Insights from the European registry of endovascular aortic repair complications (EuREC). Journal of Endovascular Surgery. 2012;**19**:37-43. DOI: 10.1583/11-3578.1

[90] Maršala M. Spinal cord blood flow and metabolism in transient spinal ischemia. In: Stålberg E, Sharma HS, Olsson Y, editors. Spinal Cord Monitoring. 1st ed. New York: Springer; 1998. pp. 5-25. DOI: 10.1007/978-3-7091-6464-8_1

[91] Uezu T, Koja K, Kuniyoshi Y, Miyagi K, Shimoji M, Arakaki K, Yamashiro S, Mabuni K, Senaha S. Blood distribution to the anterior spinal artery from each segment of intercostal and lumbar arteries. Journal of Cardiovascular Surgery (Torino). 2003;**44**:637-645

[92] De Girolami U, Zivin JA. Neuropathology of experimental spinal cord ischemia in the rabbit. Journal of Neuropathology and Experimental Neurology. 1982;**41**:129-149

[93] Boydell P. Nervous system and disorders. In: Flecknell P, editor. Manual of Rabbit Medicine and Surgery. 1st ed. Gloucester: BSAVA; 2000. pp. 57-61

[94] Zivin JA, DeGirolami U. Spinal cord infarction: A highly reproducible stroke model. Stroke. 1980;**11**:200-202. DOI: 10.1161/01.STR.11.2.200

[95] Geissler SA, Schmidt CE, Schallert T. Rodent models and behavioral outcomes of cervical spinal cord injury. Journal of Spine. 2013;**7**Suppl 4: 001. DOI: 10.4172/2165-7939.S4-001

Section 2

Macroscopical, Clinical and Surgical Anatomy

Chapter 4

Human Brain Anatomy: Prospective, Microgravity, Hemispheric Brain Specialisation and Death of a Person

Zamzuri Idris, Faruque Reza and
Jafri Malin Abdullah

Additional information is available at the end of the chapter

http://dx.doi.org/10.5772/67897

Abstract

Central nervous system seems to float inside a craniospinal space despite having miniscule amount of CSF. This buoyancy environment seems to have been existing since embryogenesis. This indicates central nervous system always need microgravity environment to function optimally. Presence of buoyancy also causes major flexure to occur at midbrain level and this deep bending area of the brain, better known as greater limbic system seems to regulate brain functions and site for cortical brainwave origin. These special features have made it as a possible site for seat of human soul and form a crucial part in discussion related to death. Besides exploring deep anatomical areas of the brain, superficial cortical areas were also studied. The brainwaves of thirteen clinical patients were analysed. Topographical, equivalent current dipoles and spectral analysis for somatosensory, motor, auditory, visual and language evoked magnetic fields were performed. Data were further analysed using matrix laboratory method for bilateral hemispheric activity and specialization. The results disclosed silent word and picture naming were bilaterally represented, but stronger responses were in the left frontal lobe and in the right parieto-temporal lobes respectively. The sensorimotor responses also showed bilateral hemispheric responses, but stronger in the contralateral hemisphere to the induced sensation or movements. For auditory-visual brainwave responses, bilateral activities were again observed, but their lateralization was mild and could be in any hemisphere. The conclusions drawn from this study are brainwaves associated with cognitive-language, sensorimotor and auditory-visual functions are represented in both hemispheres; and they are efficiently integrated via commissure systems, resulting in one hemispheric specialization. Therefore, this chapter covers superficial, integrative and deep parts of human brain anatomy with emphasis on brainwaves, brain functions, seat of human soul and death.

Keywords: microgravity, hemispheric specialization, brainwaves, soul, magnetoencephalography

1. Introduction

The average weight of the brain is 50 g in cerebrospinal fluid (CSF) and 1400 g without CSF (the actual brain weight) [1, 2]. The reduction in brain weight is believed to have resulted from the effect of CSF buoyancy or a microgravity environment created by CSF [3–5]. In principle, the force of gravity can be defied in three ways: (a) by acceleration or aerodynamic force, (b) by buoyant force that follows the Archimedes principle in 212 BC, which stated 'any object wholly or partially immersed in fluid, is buoyed up by a force equal to the weight of the fluid displaced by the object'; it is a weightlessness concept (**Figure 1A**) and (c) by an object with no (or negative) mass (? dark matter) or time (? soul). CSF buoyancy results in a reduction of actual brain weight, leading to a state of microgravity or weightlessness. An extension of this postulation is the pregnant uterus, which can exert similar effects (buoyancy resulting in microgravity). During early gestation, the ratio of foetus size to the volume of amniotic fluid is greater than the ratio at late gestation. During this period, the foetus is in a flexed position or an antigravity (microgravity) position. Therefore, it can be postulated that antigravity or microgravity environment is essential for normal development of CNS (**Figure 1B**). This stage of development leads to a flexed position of the foetus at early gestation (microgravity body position: just like the astronaut in space, curved or a horizontal position, whilst the gravity position assumes a vertical position). The microgravity position of the foetus changes at later gestation to assume a vertical gravity position, which is essential for muscles and bone development and for preparation of childbirth (with gravity force: 1g or 9.81 m/s^2) [6–8].

Regarding hemispheric specialization, the cerebrum consists of two hemispheres that are interconnected via commissures, the largest of which is the corpus callosum. Integration of information from each side appears crucial in a normal functional brain. This chapter illustrates

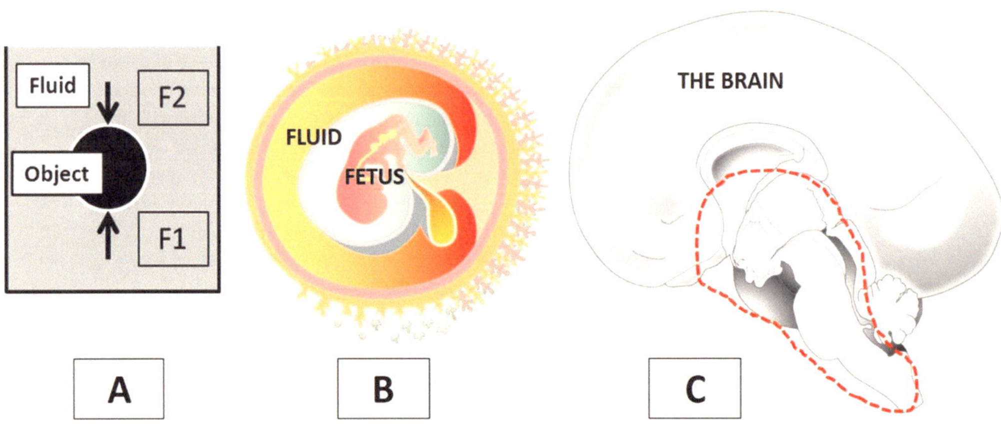

Figure 1. (A) The Archimedes principle: The sunken object will not move if the F1 force equals the F2 force. (B) Early embryogenesis which occurs in buoyant environment. The system is at its best when no influence from the gravity force is present. (C) Microgravity posture and the greater limbic area (covered by a dashed line).

the usage of magnetoencephalography (MEG) and electroencephalography (EEG) to analyse brainwaves and to map the functional anatomy of both hemispheres. Mapping and studying the functional and anatomical aspects of language, sensorimotor and auditory-visual functions have commonly been performed in other studies with positron emission tomography (PET) or functional MRI (fMRI) [9–14]. In this particular chapter, we used brainwave detection technology (MEG and EEG) to visualize the cortical brainwaves for the aforementioned tasks and study their hemispheric activity and specialization. We also performed a literature review on the anatomical structures involved in the fast and efficient transfer of information between the two cerebral hemispheres, the corpus callosum and other commissures as well as a brief review on callosal surgery.

After discussing the whole brain as an organ in microgravity environment and cortical brain anatomy and function (the superficial part of the brain),finally, in this chapter, we also discuss on the major issue related to the death of a person, which has a close anatomical link with structures at the 'deep and central part of the brain'. This deep anatomical area seems to play a crucial role in either cardiac or brain death and was labelled as 'the seat of human soul' by many ancient philosophers including Plato and Leonardo Da Vinci [15, 16]. This deep periventricular area covers anatomical structures of the brainstem, reticular system, hypothalamus, thalamus, basal forebrain or septal area, amygdala, hippocampus and pineal and pituitary glands, and it is better known as the 'greater limbic system', which was introduced by Nieuwenhuys et al. in 1988 [17, 18].

2. Microgravity inside the central nervous system

The concept of microgravity within the CNS relates to the Archimedes buoyancy effect of CSF. Despite miniscule amount of CSF, buoyancy is maintained by: (a) the Windkessel phenomenon (vascular pulsations) that causes brain pulsation and hence well-distributed intraventricular and extraventricular cerebrospinal fluid which sandwich the brain parenchyma, (b) the anchoring effect provided by the nerve roots, filum terminale, denticulate ligament at the bottom and cranial nerves as well as blood vessels at the skull base, and importantly (c) the brain itself consists of 70% of water and 30% of dry matter, and 60% of dry matter actually consists of fat. In relation to this, the proofs for the central nervous system lie in the microgravity environment and are provided by: (a) weightlessness of the brain, (b) microgravity or bending posture at the mid-brain level for the brain (therefore, terms such as ventral and dorsal, rostral and caudal for the brainstem and spinal cord and cerebrum are different: e.g., the term ventral for the brainstem is anterior whilst ventral for the cerebrum is inferior and the term rostral for the brainstem is the superior end whilst for the cerebrum, it means the anterior end) (**Figure 1C**), (c) the central nervous system development always requires buoyant environment, and this is provided by the chorionic and later by an amniotic fluid during pregnancy, (d) sinking skin flap syndrome with alteration in cerebral blood flow in a chronic craniectomy patient [19], (e) the brain seems to easily float when saline flushing is made during open brain surgery, (f) brainshift whenever CSF buoyancy is eliminated: this may suggest that the brain could indeed be in 'neutral buoyancy' by which 'CSF density' is nearly the same with 'brain density' [20, 21] and (g) studies indicating simulated microgravity enhance the differentiation of mesenchymal stem cells into neurons [8, 22]. These arguments point that the CNS could possibly lie within a microgravity environment (between 0 and 1g or 9.81 m/s^2).

In reference to aforementioned notes, this concept could explain the occurrence of flexures at the base of the brain (transitional region at the anchoring base and floating part of the telencephalon) and indicates that the thalamus and hypothalamus are possible rostral extensions of the brainstem. Furthermore, this new perspective on the CNS has several important points that should be emphasized:

a. the early development of the CNS requires microgravity environment.

b. a study of the CNS such as CNS stem cells should be done in the microgravity environment (between 0 and 1g).

c. the 'greater limbic system', as suggested by Nieuwenhuys and colleagues in 1988, is possibly a valid notion, which should include (i) the classical limbic system—amygdala, hippocampus, fornix, habenular complex, mamillary body, cingulate and parahippocampal cortices, nucleus accumbens and hypothalamus, (ii) thalamus, (iii) basal forebrain or septal nuclei, (iv) pineal and pituitary glands and (v) classical reticular-brainstem system (17, 18). This set of 'periventricular' anatomical structures should be viewed as one system, and brain networks would possibly cover at least one of its structures. This hypothesis is made based on the fact that the origin for the cortical brainwaves is from this deep anatomical area, as shown by a study done by Moruzzi and Magoun in 1949 [23].

The concepts of microgravity inside the brain, and the greater limbic system as an origin for the brainwaves that are much emphasized here, lead us to examine more on their anatomical and functional relationships.

2.1. Anatomical relationship: reticular formation network anatomy, microgravity inside the central nervous system and origin for the brainwaves

Classical reticular formation occupies the central portion of the brainstem, surrounded by the cranial nerve, sensory relay nuclei and the ascending and descending fibre systems. It is connected to all parts of the brain neocortex (six layers of cerebral hemispheric cortex), archicortex (three to four cortical layers of hippocampus and olfactory cortex) and paleocortex (four to five cortical layers of rostral insular, parahippocampus, olfactory bulb, olfactory tubercle, piriform cortex, periamygdalar area, anterior olfactory nucleus, anterior perforated substance and prepyriform area), either directly or indirectly via the basal forebrain nuclei, thalamus or hypothalamus and to the spinal cord. It is extraordinarily rich in neuromediators: noradrenalin, serotonin, choline, histamine, gamma-aminobutyric acid (GABA) and hypocretin. Generally, it can be divided into two systems: (a) ascending reticular activating system (ARAS) and (b) ascending reticular inhibitory system (ARIS) [24]. These two divisions are important in mediating consciousness, integration of autonomic (visceromotor), behavioural and somatomotor responses, the endocrine and regulation of sleep-wake cycle. The classical view of the reticular formation identifies its components only in the brainstem, with connections primarily to the thalamus, hypothalamus and basal forebrain nuclei (septal nuclei, etc.). Nieuwenhuys and colleagues provide an alternative view of the reticular system highlighting its significant involvement with the limbic, hypothalamic and parahypothalamic structures. They named this new circuit as the 'greater limbic system' and identified the hypothalamus, which resides rostrally outside the classical reticular formation as a vital component of it [17, 18].

The classical reticular formation forms diffused mosaic-like structures with many functional nuclei inside the brainstem, which includes anatomically the medulla oblongata (myelencephalon), pons (part of metencephalon) and mid-brain (mesencephalon). It forms the core of the neuroaxis, which is anatomically orientated in a vertical or gravity posture. In contrast to the brainstem, the diencephalon that consists of thalamus, epithalamus, subthalamus, hypothalamus, basal forebrain area, amygdala, hippocampus and some other periventricular structures is positioned horizontally, in an antigravity or microgravity posture. A combination of these two postures forms the 'T'-like shape of CNS cores and paracores. This is mainly resultant from the presence of mesencephalic or primary cephalic flexure during early brain development. If without this flexure, the brainstem and reticular formation shall assume a single vertical configuration with the hypothalamus-thalamus forming its rostral end. This early embryological bending occurs because of the buoyant environment provided by the chorionic and amniotic fluid during gestation and maintained throughout life by the CSF. Interestingly, a study by Moruzzi and Magoun in 1949 disclosed that the origin for the brainwaves is from this deep reticular system and influences the cortical brainwave rhythms through two pathways: (a) dorsal pathway via the thalamus (thalamocortical network) and (b) ventral pathway through the hypothalamus, basal forebrain region, amygdala and hippocampus (extra-thalamic network) [23]. This extra-thalamic network could be the reason why in refractory epilepsy, peripheral stimulation of the vagus nerve can reduce seizure rates (vagus nerve-extrathalamic pathway-hippocampus-cortex) [25–27]. These two circuits run deep inside the brain and form important circuits (core and paracore of the CNS) which deal with at least two important aspects of neurocognition: (a) consciousness and (b) memory.

2.2. Functional relationship: consciousness, memory and origin for the brainwaves

Consciousness and memory are seen as two essential aspects in human cognition. This mental process of acquiring knowledge and understanding through thought, experience and senses is special for human beings. This cognitive capability also allows some humans who are believers to appreciate creations and God (creator). One may find difficulty in praying to God if he or she had an alteration in conscious or memory level. Therefore, one may view that these cores and paracores of the CNS which give rise to consciousness and memory are essentially a seat of human soul. The debates on the seat of human soul had been going on since ancient times. Plato (424–348 BC) and Galen (circa 200/216 BC) had labelled the brain (encephalocentric theory) as an important organ for the soul whilst Aristotle (circa 384–383 BC) who learnt from Plato disagreed on Plato's idea and preferred the heart as the seat of human soul. Later, during the renaissance period, which began roughly at fourteenth century in Italy, Leonardo da Vinci (1452–1519) had located the soul inside the brain and more specifically in the middle ventricle close to anterior portion of the third ventricle near the hypothalamus after drawing the intersecting infinity lines (golden ratio) of the human cranium [5, 15, 16]. The area identified by Leonardo da Vinci is in fact part of the greater limbic system [interesting to note that most structures in this deep anatomical area are infinity in shapes—such as Solomon's knot (mosaic-like reticular system), Pascal's spiral, Archimedean and Durer spirals (hippocampus, caudate nucleus), cycles of Lemniscate (thalamus) and pyramid (insular)]. Therefore, it seems that the greater limbic system is an attractive notion for the seat of human soul because of several reasons:

a. It is an area for 'brainwave origin'.

b. It controls 'consciousness and memory' (two main aspects of human cognition and closely related to remembering God); alteration or loss of consciousness (or memory) happens if someone injured this deep area and therefore have difficulty to remember or appreciate God.

c. A person's death would involve this anatomical area—refer to the last section in this chapter.

d. It may be viewed as the centre of 'all brain networks' (at least one node which arises from this deep brain region may be present in any brain network, and this node could appear larger than the rest).

e. 'Infinity' lines of the skull intercepting at this area, and most anatomical structures in this deep area, are likely 'infinite' in their shapes.

We have discussed the whole brain and viewed it as one in microgravity environment and touched on the curving region of the brain (periventricular region or deep region of the brain), which forms a core and a paracore of the CNS that regulates brainwave rhythms, controls consciousness and memory and determines death of a person. Before discussing further on matters pertaining to death of a person and deep brain area, next we present our study on neurocognition, which commonly involves the superficial brain area or two cortical brain hemispheres, which is also known as 'bilateral hemispheric involvement and hemispheric specialization'.

3. A study on hemispheric human brain specialization

Cortical brainwaves mainly result from pyramidal postsynaptic potentials, which have synchronized oscillations with the following: (a) the thalamus, otherwise known as thalamocortical networks, which are modulated by the reticulo-thalamo-cortical circuits, (b) the extrathalamic-cortical circuits, which mainly involve the reticular system, hypothalamus, hippocampus, amygdala, basal forebrain and septal nuclei and (c) other cortices, known as cortical-cortical networks [5, 17]. In 1952, Magoun reported that the reticular system in the brainstem has a crucial role in generating the pattern of brainwaves [23]. This classical reticular system in the brainstem has vast networks with other structures in the diencephalon, such as the thalamus, hypothalamus, basal forebrain and septal nuclei, parahypothalamic nuclei, pineal and pituitary glands, the limbic system as well as the insula, basal ganglia and neocortex. The vast interconnecting networks, via the thalamic and extrathalamic circuits, create optimal brainwave oscillations in the cortex, which can be studied using MEG and EEG [4, 28].

Generally, it is complicated to map the actual areas responsible for brain cognition, sensorimotor and auditory-visual functions. Many believe that these brain functions could have originated deep within the centre of the brain, involving anatomical areas that have vast networks with the cortices [28–32]. These areas are the thalamus, hypothalamus, amygdala, hippocampus, basal forebrain and septal nuclei, reticular system and pituitary-pineal system which form the core and paracore for the central nervous system. Mapping the areas involved in the aforementioned functions should ideally have covered these deep areas. However, our study focused only on superficial brain mapping and cortical brainwave analysis as the availability of MEG testing allows relatively reliable, superficial and non-invasive methods compared to deep brain mapping [33].

3.1. Studied subjects

This chapter included 13 clinical, adult, right-handed patients with various pathologies as follows: cortical dysplasia, meningioma, low- and high-grade gliomas, glioblastomas (GBM), basal ganglia arteriovenous malformation, temporal arteriovenous malformation, cavernomas and periatrial lesions (**Table 1**). All patients underwent routine MEG recordings before the neurosurgical interventions. MEG recordings were made for standard evoked somatosensory, motor, auditory and visual responses. For patients with lesions near the assumed speech area, further language MEG recordings and mappings were performed. The MEG data were registered, processed and fused with anatomical MRI images. These images were then used with the neuronavigation system for surgery. Two patients underwent contralateral hemispheric scalp EEG recordings during awake brain surgery (cases 1 and 2 in **Table 1**).

3.2. MEG recording, procedure, post-processing and overdetermined anatomical analysis for somatosensory-, motor-, auditory- and visual-evoked fields

Magnetic-evoked fields were recorded whilst patients were seated in a magnetically shielded room (MaxShieldTM, ElektaOy, Helsinki, Finland) using a 306-channel (102 magnetometers and 204 gradiometers) whole-head MEG system (ElektaNeuromag®, ElektaOy, Helsinki, Finland) (**Figure 2A**). Online band-pass filtering was performed between 0.01 and 330 Hz to discard the noise. Further filtering was performed for offline data analysis using a high-pass filter of 60 Hz with a width of 0.6 Hz and a low-pass filter of 3 Hz with a width of 0.3 Hz. The epoch duration was up to 300 ms. The sampling frequency was 1 kHz. With respect to the procedure, the head position relative to the MEG sensors of the helmet was localized using

Types of analysis	Diagnosis	MEG/EEG analysis	MEG/EEG findings	Summary of the responses
Motor responses	1. Motor cortex metastases	a. Topographical non-superimposed MEG motor-evoked fields and	a. MEG: Bilateral-evoked fields with stronger fields noted contralateral to the movements	Bilateral hemispheric motor responses with stronger fields in the contralateral side to the movements and inversed polarity in the hemisphere that was ipsilateral to the movements
		b. Contralateral to the stimulated motor gyrus EEG analysis (during awake surgery)	b. EEG: Contralateral responses were recorded when the motor gyrus was stimulated	
	2. Right periatrial lesion	a. Topographical non-superimposed MEG motor-evoked fields and	a. MEG: Bilateral-evoked fields with stronger fields noted contralateral to the movements	
		b. Contralateral to the stimulated motor gyrus EEG analysis (during awake surgery)	b. EEG: Contralateral responses were recorded when the motor gyrus was stimulated	
	3. Basal ganglia vascular lesion (AVM)	a. Topographical superimposed MEG motor-evoked fields	a. MEG: Bilateral hemispheric motor-evoked fields responses with stronger fields in the hemisphere contralateral to the movements	
	4. Motor cortical dysplasia	a. Topographical non-superimposed bilateral MEG motor-evoked fields	a. MEG: Bilateral hemispheric motor-evoked field responses with stronger fields in the hemisphere contralateral to the movements	
Sensory responses	1. Left lower frontal low-grade gliomas	a. Brain lobes MEG analysis for sensory-evoked fields	a. Bilateral hemispheric sensory-evoked field responses that were stronger in the hemisphere that was contralateral to the sensory median nerve stimulation	Bilateral hemispheric sensory responses were stronger in the hemisphere that was contralateral to the sensory stimulation
	2. Right frontal-temporal glioblastomas (GBM)	a. MEG source analysis for sensory-evoked fields	a. Bilateral hemispheric sensory-evoked field responses that were stronger in the hemisphere that was contralateral to the sensory median nerve stimulation	

Types of analysis	Diagnosis	MEG/EEG analysis	MEG/EEG findings	Summary of the responses
	3. Right temporal meningioma	a. MEG source analysis for sensory-evoked fields	a. Bilateral hemispheric sensory-evoked field responses that were stronger in the hemisphere that was contralateral to the sensory median nerve stimulation	
Auditory responses	1. Cribriform plate meningioma	a. Brain lobes and topographical MEG analysis for auditory-evoked fields	a. Bilateral hemispheric auditory-evoked field responses with mild dominance in the ipsilateral hemisphere to ear clicks for hearing	Bilateral hemispheric auditory responses with mild dominance in either hemisphere
	2. Small basal ganglia vascular lesion (AVM)	a. Brain lobes and topographical MEG analysis for auditory-evoked fields	a. Bilateral hemispheric auditory-evoked field responses with mild dominance in the contralateral hemisphere to ear clicks for hearing	
	3. Left frontal-temporal meningioma	a. MEG source analysis for auditory-evoked fields	a. Bilateral hemispheric sensory-evoked field responses. Source localization and brain activity of the auditory area are matched. Stronger activation on the left hemisphere	
Visual responses	1. Right periatrial lesion	a. Topographical non-superimposed MEG visual-evoked fields and Matlab analysis	a. Bilateral visual-evoked field responses with mild left dominance	Bilateral hemispheric visual responses with mild dominance in either hemisphere
	2. Right parietal cavernoma	a. Topographical non-superimposed MEG visual-evoked fields	a. Bilateral visual-evoked field responses with mild right dominance	
	3. Right temporal arteriovenous malformation	a. MEG source analysis for visual evoked fields	a. Bilateral hemispheric visual-evoked fields responses. Source localization and brain activity are matched. Brain activation of both hemispheres is recruited equally	
Language responses—silent word naming	1. Left lower frontal low-grade gliomas	a. Brain lobes MEG and Matlab analysis for silent word naming	a. Bilateral frontal, parietal, temporal and occipital lobe responses with marked differences (stronger) in the left frontal lobe (dominant hemisphere)	Bilateral hemispheric responses with dominance in the left hemisphere

Types of analysis	Diagnosis	MEG/EEG analysis	MEG/EEG findings	Summary of the responses
	2. Left upper frontal low-grade tumour	a. MEG source analysis for silent word naming of language-evoked field	a. Source analysis indicated that there is bilateral source localization in the temporal area. Source localization and brain activity are matched. Stronger activation on the left hemisphere	
Language responses—silent picture naming	1. Left lower frontal low-grade gliomas	a. Brain lobe MEG analysis for silent picture naming	a. Bilateral parietal, temporal and occipital lobe responses with stronger responses noted in the right temporal lobe	Bilateral hemispheric responses with stronger responses noted in the non-dominant (right) temporal and parietal lobes
	2. Left temporal high-grade gliomas	a. Brain lobe MEG and Matlab analysis for silent picture naming	a. Bilateral responses with stronger responses noted in the right temporal lobe	

Table 1. Summary of clinical cases that were studied using MEG and EEG for hemispheric specialization.

the following: (a) three fiducial localization coils attached to the right and left pre-auricular points and to the nasion of the patient, (b) 100–150 points digitized around the head using a 3D position monitoring system (Pholemus, Colchester, VT) and (c) four electromagnetic head position indicator (HPI) coils to assess the head position at the beginning of the measurement process. During the recording, head position changes of up to 1.5 cm were accepted. MEG source localization for somatosensory- (stimulation of the median nerve in the hand), motor- (active movement of the index finger), auditory- (emission of clicking sounds in each ear separately) and visual (each eye tested with a checkerboard separately)-evoked magnetic fields were performed using the overdetermined equivalent current dipole (ECD) technique, which was already installed inside the Neuromag computer working station (**Figure 2B**). The somatosensory-, motor-, auditory- and visual-evoked magnetic fields for a person without intracranial pathology is expected to be at around N20 (20 ms), P5 (−5 ms) (left-hand motor), P50 (−50 ms) (right-hand motor), N100 (100 ms) and N75–120 (75–120 ms), respectively (N: negativity and P: positivity). The anatomical magnetic resonance imaging (MRI) of T1, T2, FLAIR and 3D sequences were obtained using Philips MRI (Philips Intera 3.0T MRI scanner). Fusion between the anatomical MRI images and topographic reconstruction of the head-model brainwave data was completed prior to source localization.

3.3. MEG recording, procedure, post-processing and overdetermined anatomical analysis for language

The MEG equipment, software and sampling rates were the same as the one described above using an Elekta MEG-Neuromag Ltd, with 306 channels consisting of 204 planar gradiometers and 102 magnetometers, which were set at a minimum sampling rate of 1 kHz. The band-pass filter was between 0.01 and 330 Hz, with a high-pass filter of 60 Hz and width of 0.6 Hz and a low-pass filter of 3 Hz, with a width 0.3 Hz. The epoch duration for the language study was longer (850 ms), including a −150 ms pre-stimulus interval. Silent reading tasks were performed during MEG recordings, where subjects sat on a comfortable chair with their heads fixed into the MEG machine. After the presentation of an eye fixation point for 3 s, four-character semantic words for the word-naming task were shown for 3 s on an 80-inch rear projection screen that was located 1.5 m away from the subject in the same room. Visual stimuli were generated using a visual presentation system which was projected by a projector located outside the room. Subjects were tasked to read immediately after the presentation of the word only once, without phonation. One session consisted of 100 different word presentations. The words were selected from an elementary school dictionary so that the subjects would quickly and easily understand them. The word stimuli subtended a horizontal visual angle of 3° and a vertical angle of 1°; as a result, no eye movements were necessary to visualize the presented word. Each recording session took at least 1 h to complete; however, the subjects were able to pause the task if they were starting to feel uncomfortable. The same procedure was repeated for picture naming, whereby common pictures were shown and patients silently named the pictures. The analysed brainwave language-related field (LRF) components included N100 (100 ms), N200 (200 ms), N400 (400 ms) and N600 (600 ms). The components were taken from the highest peak of each evoked LRF signal. The evoked LRF data were analysed in topographical brain lobes, then were fused with the anatomical MRI images and further subjected to the underdetermined modelling analysis using Matlab-statistical parametric mapping (SPM) and brain electrical source analysis (BESA) software.

3.4. Underdetermined anatomical analysis for MEG data

An in-house Matlab-SPM-based MEG-pipeline programme was used to analyse the MEG data. This was accomplished with SPM-based Matlab 7.4–R2008a (MathWorks Inc., Natick, MA, USA) to diffusely localize eloquent areas based on Montreal Neurological Institute (MNI) template. Standard neuroscience spectral data analysis, such as analysis on the region of interest (ROI) with the concomitant detection of significant active regions ($p < 0.05$) that respond to external stimuli and inverse solutions for EEG or MEG data, was utilized (**Figure 2C–E**). Besides an in-house SPM-based Matlab, BESA (Version 6.0, GmbH, Graefelfing, Germany) was also used to process the source localization of the waveforms for the sensory, visual, auditory and language processing area. MEG data were co-registered to the template of structural MRI implemented in BESA Research 6.0. Two source dipoles were fitted with the constraint of having symmetrical sources in each hemisphere. Using different start locations, these symmetric dipoles were allocated consistently to the region of interest. Dipoles were fitted sequentially; a single dipole was placed on the right hemisphere and fitted over 50–150 ms for auditory-, visual- and language-evoked responses and 0–50 ms for somatosensory-evoked responses. These steps were subsequently repeated in the opposite hemisphere.

3.5. Results on data analysis

MEG data of 13 clinical patients were analysed. This included two patients who had scalp electrodes on the opposite hemisphere and direct motor cortex stimulation during awake

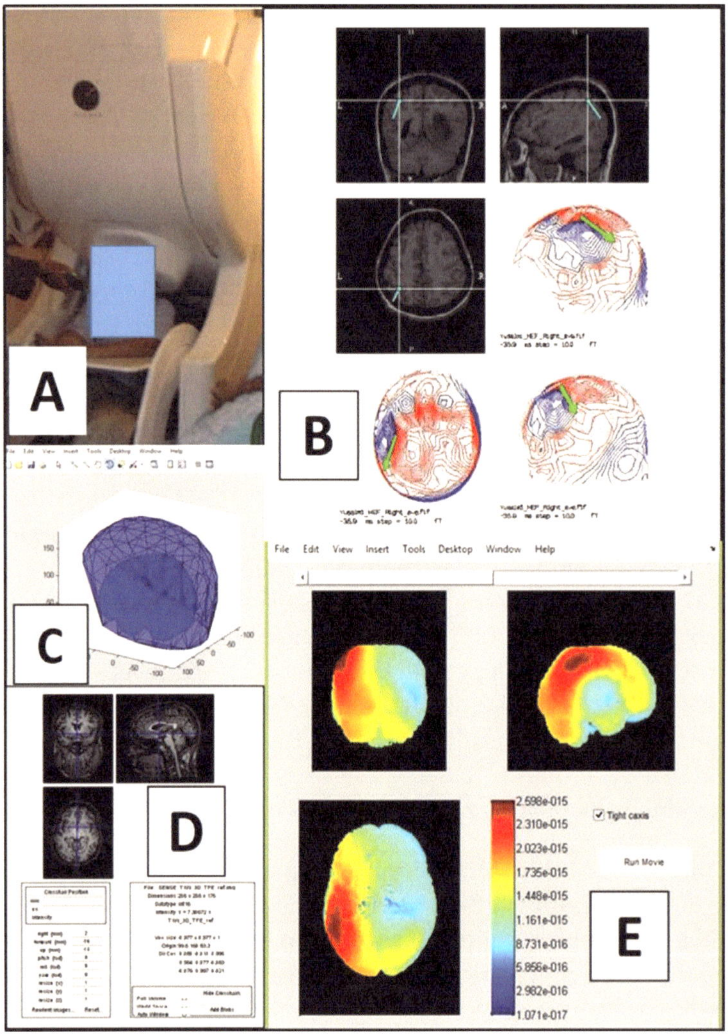

Figure 2. (A) MEG recording procedure. (B) An overdetermined equivalent current dipoles analysis using the vector of magnetic fields to localize the source of various evoked fields. (C) Spherical modelling for the brain using Matlab. (D) Image fusion between MEG-brainwave and MRI brain. (E) Underdetermined inverse solutions using Matlab for MEG data that have a significant response to the evoked magnetic field (in this example, it is the somatosensory-evoked field).

brain surgery. All patients underwent MEG prior to any surgical intervention for the purpose of mapping the eloquent anatomical areas of the brain. The MEG data were analysed for motor-, sensory-, auditory-, visual- and language-evoked fields. The summary of the analysis is presented in **Table 1** (17 analyses from 13 patients).

3.5.1. Hemispheric responses for motor-, sensory-, auditory- and visual-evoked fields

Unilateral motor-, sensory-, auditory- and visual-evoked fields were present in both cerebral hemispheres. There were some peculiar differences amongst them. For motor-evoked fields, there were bilateral hemispheric responses with stronger responses from the hemisphere

http://dx.doi.org/10.5772/67897

contralateral to the finger movement (**Figure 3A** and **B**). Two of the four patients analysed for motor responses underwent awake surgery with direct motor cortex stimulation and contralateral scalp EEG monitoring. The scalp EEG recordings demonstrated inverse polarities produced by unilateral hand movements where upgoing waveforms were seen in the contralateral hemisphere and downgoing waveforms were seen in the ipsilateral hemisphere (**Figure 3C–E**). These inversed polarities were further confirmed with topographical MEG brainwave analysis for motor functions as shown in **Figure 3A** and **B**. Similarly, sensory-evoked fields were studied using MEG, and responses were noted in both hemispheres with markedly stronger responses observed in the hemisphere contralateral to the sensory stimulation (**Figure 4**). Results from source localization and brain activation analysis of the other two patients also showed a similar pattern of responses, bilateral activities and a stronger activation on the contralateral sensory areas (**Figure 4C** and **D**).

For auditory-evoked fields, three patients were included in the analysis. Source localization and brain activation results were matched and demonstrated with bilateral activation of the

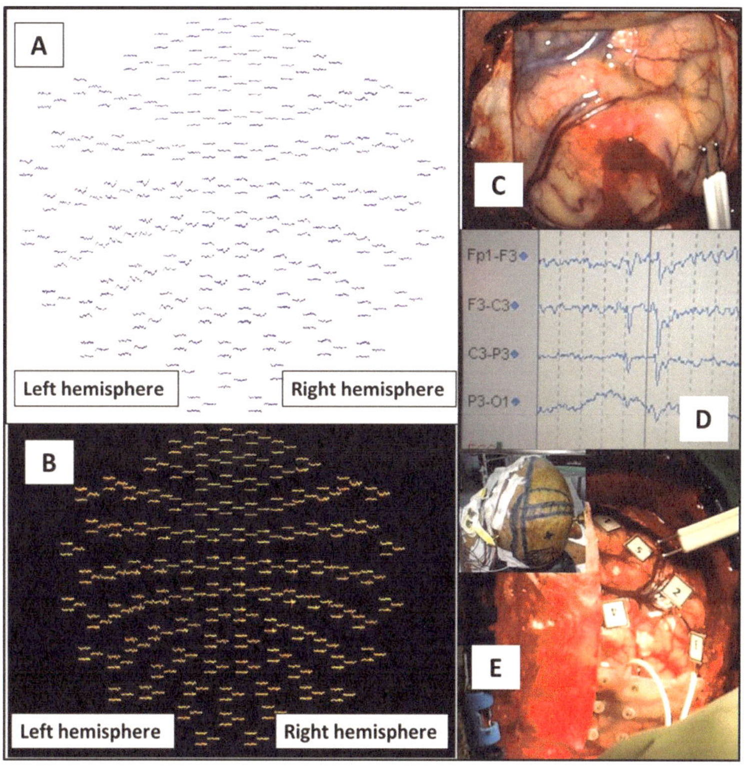

Figure 3. (A) Topographical non-superimposed motor-evoked fields for a right-finger tap. Bilateral hemispheric motor-evoked fields responses but stronger in the hemisphere contralateral to the movements (left hemisphere). (B) Topographical superimposed motor-evoked fields for the right-finger tap (red-evoked fields) and left-finger tap (yellow-evoked fields). Bilateral hemispheric motor-evoked fields responses are again noted here but a stronger response is seen in the hemisphere contralateral to the movements. (C and D) Direct motor cortex stimulation induces inversed-spike waves polarity in the contralateral hemisphere detected by scalp EEG. (E) A similar procedure in another patient showing similar findings (inset shows the scalp EEG in a contralateral hemisphere).

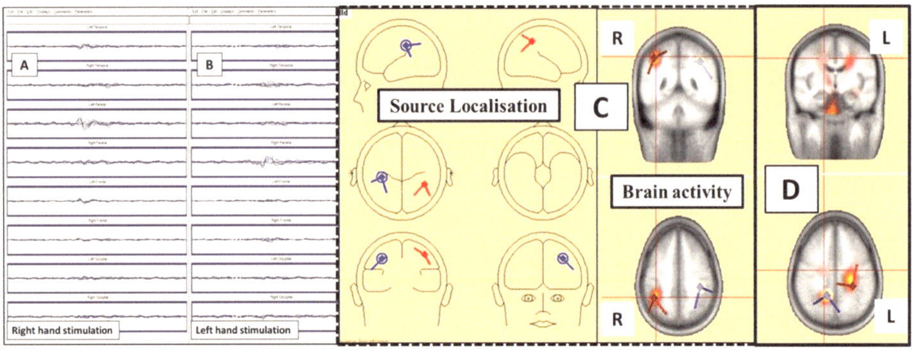

Figure 4. (A) Topographical brain lobe analysis: Right median nerve stimulation-induced bilateral hemispheric responses but a stronger response is noted in the contralateral hemisphere. (B) Similar findings were noted when the left median nerve was stimulated. (C and D) BESA-based source and brain activity analysis of another two patients during sensory-evoked field responses. Results indicate bilateral activations in the sensory areas but with a stronger activation in the contralateral hemisphere to the sensory stimulation (C: left-hand sensory stimulation and D: right-hand sensory stimulation).

auditory areas with mild hemispheric specialization. Moreover, the hemispheric dominance for auditory responses was noted as non-specific; it can either be in the right or left hemisphere (**Figure 5A–E**). This was because the waveforms produced by auditory stimulation were nearly similar in both hemispheres and indicated that auditory dominancy was indeed mild. For visual-evoked fields, there were again nearly similar bilateral brainwave representations and, therefore, unclear hemispheric specialization was observed on topographical images. As before, advanced source analysis and brain activation results again confirmed bilateral activations in the visual areas with mild hemispheric specialization. **Figure 6A** and **B** shows bilateral activations with mild hemispheric specialization in the left whilst in **Figure 6C**, the right side is the dominant hemisphere for visual-evoked fields.

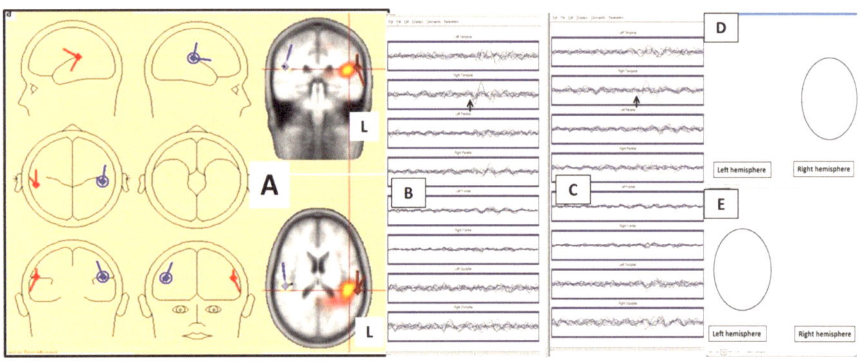

Figure 5. (A) Source localization and brain activity in the bilateral temporal areas during auditory-evoked field responses analysed using BESA. Results show stronger responses in the left hemisphere. (B and C) Topographical brain lobe analysis. (B) Right ear clicks resulting in auditory-evoked magnetic fields in ipsilateral ear (arrow head shows earlier responses in right temporal with smoother and well-formed waveforms). (C) A similar patient with left ear clicks, resulting in auditory-evoked magnetic field responses in the contralateral ear (arrow head shows earlier responses also in a right temporal with smoother and well-formed waveforms). Therefore, the right hemisphere is mildly dominant for hearing in this patient. (D) A similar patient in B and C with its data portrayed in topographical brainwave analysis for hearing—left ear clicks induced right hemispheric responses (circle). (E) Another patient who had left ear clicks but main responses were seen in the left hemisphere (circle); the subject has mild ipsilateral or the left hemisphere is dominant for hearing.

http://dx.doi.org/10.5772/67897

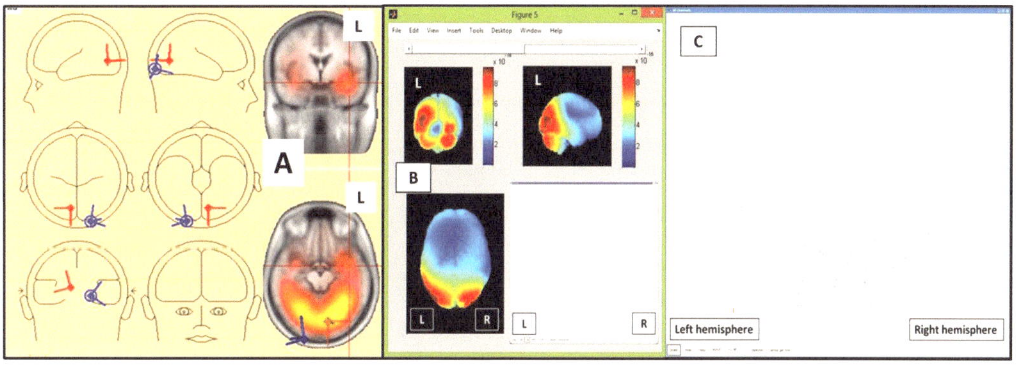

Figure 6. Source localization and brain activity for vision. (A) BESA-based analysis shows activation of the bilateral occipital areas during visual-evoked fields with mild hemispheric dominance in the left. (B) Topographical visual-evoked fields and Matlab-SPM-based analysis show bilateral visual-evoked field responses with mild left dominancy. (C) Topographical visual-evoked fields of another patient depicting bilateral visual-evoked field responses but with mild right dominancy.

3.5.2. Hemispheric responses for language—silent word and picture naming

Brainwave analysis for language study also showed bilateral hemispheric responses. For silent word naming, brainwave activities were more markedly noted in the left than in the right frontal lobe, which could reflect the Broca's speech area (**Figure 7A**). This magnetic-evoked field, which was localized over the left frontotemporal area, was subsequently confirmed during awake brain surgery (**Figure 7B** and **C**). The Matlab-SPM-based analysis for silent word naming also revealed bilateral hemispheric responses that were more pronounced in the left hemisphere (**Figure 7D**). Nonetheless, one must be reminded that the right frontal lobe may also be involved in speech. Similarly, for silent picture naming, the activities were also bilateral but more was noted in the right temporal and parietal lobes as depicted on brainwave topographical brain lobe images, magnetic-evoked fields and Matlab-SPM-based diffused underdetermined methods (**Figure 8**).

3.6. Discussion: bilateral hemispheric responses and hemispheric specialization for motor, sensory, auditory, visual and language

Cutting-edge clinical neuroimaging of MEG and EEG enables the study of brain activity as images (brainwaves) and depicts functional networks of the brain. This study showed that not only does language have a feature of hemispheric dominance, as shown by Pierre Paul Broca in 1861 [34] but also hemispheric dominance for motor, sensory, auditory and visual cortical functions. Hemispheric specialization or dominance is defined as a hemisphere-dependent relationship between a specific function and a set of brain structures, which includes both hemispheric interaction by a given hemisphere of specialized networks that have unique functional properties and its mechanisms, enabling efficient interhemispheric coordination [35]. This functional lateralization or dominance is related to the grey and white matter asymmetries, which are established early in life, and directly suggests a strong relationship with the underlying genetic factors, as noted in various studies on functional MRI and diffusion tensor imaging [36–39]. Our study is different from previously published studies as

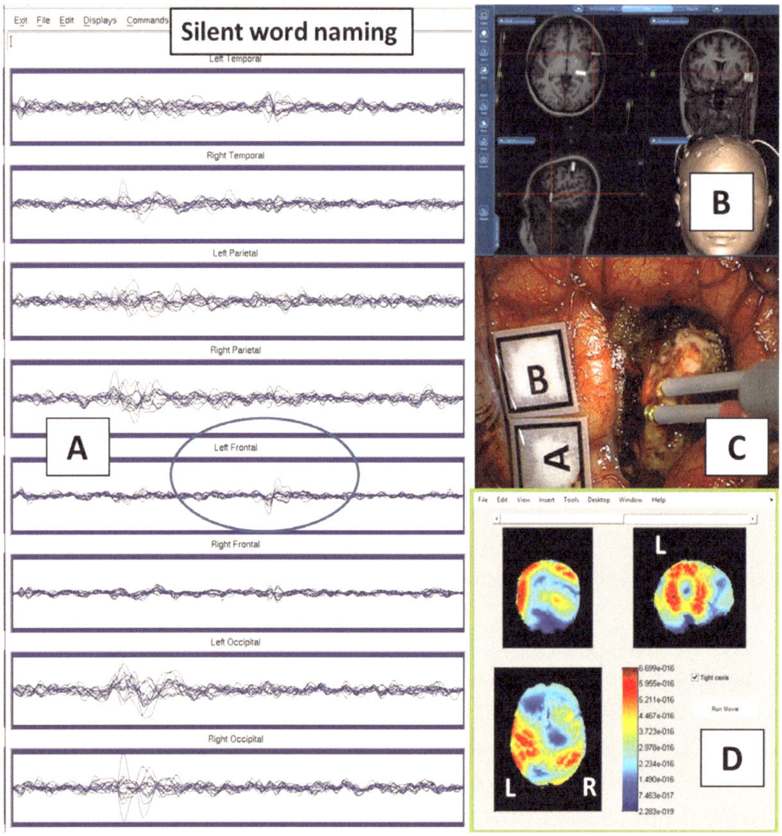

Figure 7. (A) Brainwaves analysis for silent word naming shows bilateral hemispheric responses; the activities were more marked in the left frontal lobe (circle) as compared to the right frontal which could reflect the Broca's speech area. (B and C) This magnetic-evoked field for silent word naming which was localized over the left frontotemporal area was subsequently confirmed during awake brain surgery. B image shows the navigation system during surgery which localized the speech area and was confirmed with direct brain stimulation as shown in image C. (D) The Matlab analysis for silent word naming also revealed bilateral hemispheric responses but more pronounced responses in the left hemisphere.

we used brainwaves (MEG and EEG) as the main parameter to study hemispheric activity and hemispheric specialization (dominance) for various tasks. Our brainwave study supports the findings of previous studies on hemispheric specialization using various other modalities [9, 10, 40–45]. Our chapter highlights that for sensorimotor activity, marked brainwave responses were noted in both hemispheres with a preference (lateralization or dominance or specialization) for one hemisphere. The motor brainwave responses were bilateral, and stronger wave responses were definitely noted in the hemisphere that was contralateral to the movements. The hemisphere that was ipsilateral to the movements was also activated, but it had inversed brainwave polarities. This suggests that integration of information from both hemispheres plays an essential role in carrying out efficient sensorimotor functions. By contrast, the results for hemispheric specialization for auditory and visual functions were unpredictable, and either hemisphere could be dominant (non-fixed). This could possibly be because lesions were present or because of a genetic factor that determines which hemisphere

http://dx.doi.org/10.5772/67897

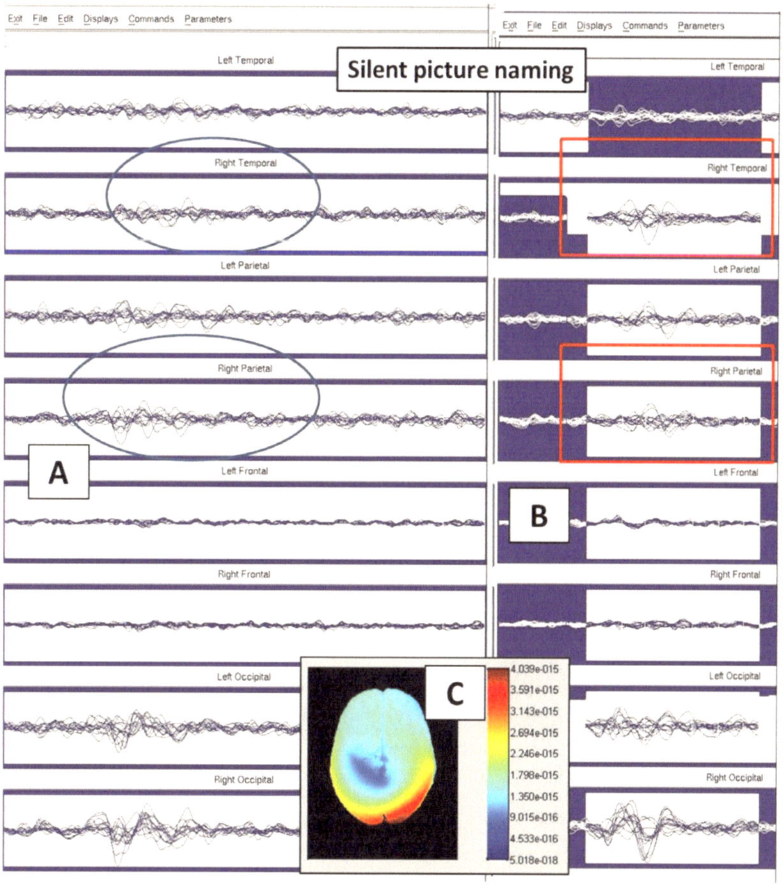

Figure 8. (A and B) Topographical brain lobe analysis for silent picture naming in two separate patients. The activities were also bilateral, but stronger activities were noted in the right temporal and parietal lobes (circles and rectangles). (C) Matlab-diffused underdetermined analysis confirmed the right shift.

is the dominant hemisphere for both auditory and visual functions. The genetic factor is the more likely explanation here as we had one patient with a mid-line lesion who underwent a hearing assessment and two patients with a right-sided lesion who had a visual assessment, and analysis of their data showed that either hemisphere could be a dominant hemisphere for auditory and visual functions. In addition, it is worth noting that there was only mild hemispheric specialization for both auditory and visual responses. In this respect, one cannot simply label auditory dominance based on the side of the ear that is commonly used for the telephone. This particular feature may arise because of the handedness of the person rather than the dominant character of auditory cortex. For cognitive-language brainwave responses, bilateral hemispheric responses were also noted. However, for silent word naming, there were more marked responses arising from the left frontal lobe in right-handed patients which suggest that silent word naming lateralizes to the dominant or left hemisphere. On the other hand, the brainwave study for silent picture naming in two right-handed subjects lateralized to the right hemisphere as there were more marked responses in the right parietal and temporal lobes. This indicates that hemispheric lateralization for visuospatial attention is in the right hemisphere, which is in agreement with findings from other studies [46–48]. Although there has been progress in elucidating the neural basis of right hemispheric dominance for this function, there is little evidence supporting its origin. One theory considers right hemispheric specialization for certain tasks as a side-effect or overload of left hemisphere dominance for language, whereas another theory considers that this division of hemispheric specialization is a reflection of the genetic, biological or environmental conditions or a combination of these [35, 49]. In conclusion, both brain hemispheres are necessary to integrate information for cognition, sensorimotor and auditory-visual functions, but there is stronger lateralization or specialization (dominance) for sensorimotor and language functions and mild for auditory and visual specialization in one hemisphere. The need for information integration by bilateral hemispheres results in specialization (dominance or lateralization) of the hemisphere. This information-integration process in the form of brainwaves is accomplished by axonal connections between the two cerebral hemispheres which are well known as commissures. The largest of these is the corpus callosum (noteworthy that significant contribution can be made further by mathematicians in elucidating this integration process).

4. A review on corpus callosum, callosal surgery and commissures

The corpus callosum is a broad, transverse bundle of myelinated nerve fibres connecting the right and left cerebral hemispheres (**Figure 9A**). Anatomically, it is divided into the following five regions: rostrum, genu, body, isthmus and splenium. It has been suggested that such a connection and anatomical division are modality-specific; the anterior callosal fibres interconnecting the frontal lobes transfer motor information and the posterior fibres connecting the parietal, temporal and occipital lobes bilaterally are responsible for the integration of somatosensory (posterior mid-body), auditory (isthmus) and visual (splenium) information [50, 51]. Embryologically, the corpus callosum forms in an anterior to posterior direction with the genu forming first, followed by the body, isthmus (marked with a slight narrowing at the level where the fornix abuts the callosum), splenium and rostrum [51–54]. It develops from the upper segment of the telencephalic alar plate via the following four stages: (a) prosencephalic cleavage (28–35 days of gestation), (b) commissural plate formation (36–73 days of gestation), (c) corpus callosum formation (74–115 days of gestation) and (d) corpus callosum growth (after 115 days of gestation). During the prosencepalic cleavage period, the prosencephalon splits into the telencephalon and diencephalon. Subsequently, the single telencephalon leads to the formation of two telencephalic vesicles and a floor between them, which is called the lamina terminalis. During the commissural plate formation period, the lamina terminalis thickens and is called the lamina reuniens or commissural plate. The commissural plate continues to thicken, and by 73 days, the following four structures can be appreciated within it: (a) the site of the future corpus callosum, (b) area of the future anterior commissure, (c) hippocampal commissure and (d) septum cavum pellucidum. From 74 days onwards, the corpus callosum is formed from the crossed cortical axons through the area of the commissural plate. The axons from different regions of the brain cross at 'different times', resulting in different regions and functions of the corpus callosum (**Figure 9B**). In contrast to corpus callosum formation, the maturation and myelination process starts from the posterior to anterior [55, 56]. It begins to appear postnatally

http://dx.doi.org/10.5772/67897

in the splenium by approximately 4 months and in the genu by approximately 6 months. The corpus callosum has an adult appearance by approximately 8 months of age and continues to develop through the first two decades of life by a progressive increase in its size [57, 58]. These myelinated axons permit the fast propagation of neural impulses or waves that are considered prerequisites for normal cognitive, sensorimotor and auditory-visual functions. Indeed, abnormalities in the corpus callosum, especially those with associated brain anomalies and syndromic types of agenesis, are correlated with impairment in neurocognition, neurobehavioural, sensorimotor and auditory-visual functions [59–62]. These lines of suggestion indicate that the corpus callosum is a vital structure for cortical-cortical and interhemispheric connectivity, reflecting a computational requirement of interhemispheric coordination for normal behaviour, cognition, sensorimotor and auditory-visual functions.

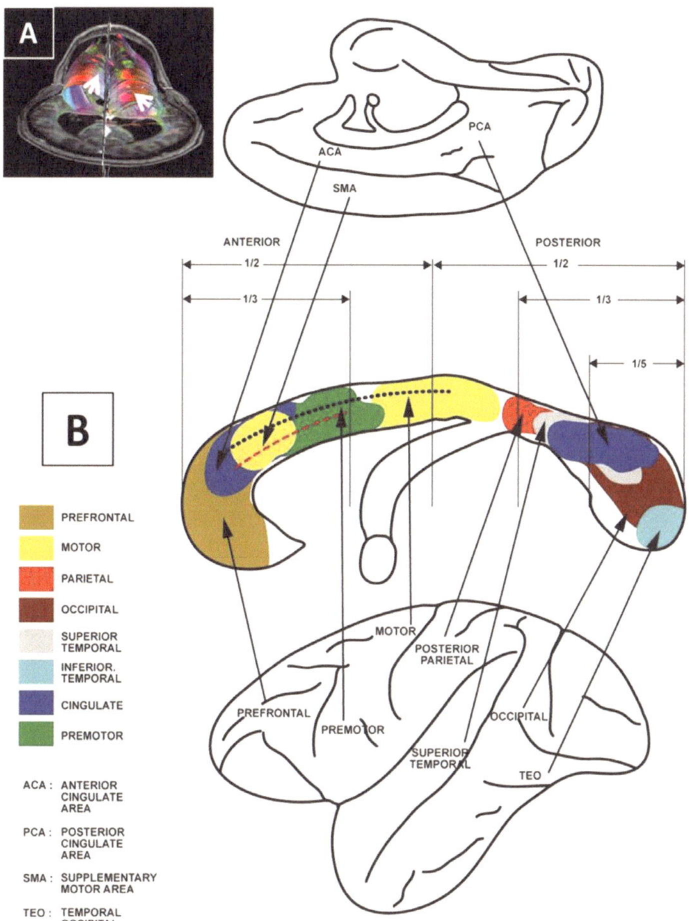

Figure 9. (A) Corpus callosum is a broad transverse bundle of myelinated nerve fibres, connecting the two cerebral hemispheres as shown here on the fibre tracking image (two white arrow heads). (B) Depicting crossing fibres from different regions of the brain passing through the corpus callosum and sites for surgical intervention: black dots—callosotomy site for refractory epileptic akinetic seizures—and red dashes—callosotomy for the surgical approach to lateral and third ventricles.

With respect to callosal surgery, it should be performed carefully, with adequate background knowledge on its anatomy and connectivity. The anterior interhemispheric transcallosal approach to the lateral and/or third ventricles should resect the anterior part alone, the rostral body and part of the genu, sparing the crossing motor fibres from the primary motor cortices in the anterior mid-body and, hence, avoiding motor complications [63] (**Figure 9B**). The posterior interhemsipheric transcallosal approach is rarely used to reach the pineal region and posterior part of the third ventricle. This approach involves resection of the splenium, which may cause somatosensory, auditory, visual or emotional disturbances. Some patients may appear grossly intact and unchanged when observed by family and friends, but when specific neuropsychological tests are administered after the surgery, the deficits can be significant. Some examples of these deficits are verbal anosmia, double hemianopsia, poor processing of verbal information, apraxia or agraphia of the left hand. By contrast, resective callosotomy for intractable epilepsy due to severe, medically intractable seizures, where akinetic seizures or drop attacks are a predominant feature, will respond favourably to corpus callosum resection [64, 65]. Callosal division should be performed as described above. Resection can be extended further anteriorly until the rostrum, where the anterior commissure is an anterior limit and is best appreciated when seeing the two fornices converge together (**Figure 9B**). The resection should ideally be extended posteriorly to cover the anterior, two-thirds of the corpus callosum, especially in cases where the seizure outcome is unsatisfactory. This means that resection should include the motor fibres that run in the anterior and, possibly, part of the posterior mid-body, which carry the risk of permanent motor deficits. Hence, the posterior limit is more difficult to estimate and is commonly guided by the expected clinical outcomes (objective of the surgery), navigation system, thinning of the body (isthmus) and appearances of the fornices (the isthmus is the area where the fornix abuts the corpus callosum).

Other known commissures that cross the mid-line, connecting the two cerebral hemispheres, are the anterior, hippocampal or forniceal, habenular, posterior or epithalamic and supraoptic commissures [53, 54]. The anterior commissure can be found on either side, beneath the corpus striatum and in the substance of the temporal lobe. It connects the two amygdala and temporal lobes and contains decussating fibres from the olfactory tracts. It is part of the neospinothalamic tract for pain. The hippocampal or forniceal commissure is the second largest of the commissural connecting bundles that join the two crura of the fornix and connect the two hippocampi. Next is the habenular commissure, which is situated in front of the pineal gland and connects the habenular nuclei on both sides of the diencephalon. It has connections with the pineal and interpeduncular nuclei in the mid-brain. The second to last is the posterior commissure, which is a rounded band of white fibres crossing the mid-line on the dorsal aspect of the upper end of the cerebral aqueduct. It interconnects the pretectal nuclei and mediates the bilateral pupillary light reflex. Finally, the supraoptic commissure or decussation is the crossover within the optic pathway system, which interconnects the two eyes with the two visual cortices. Anatomical knowledge of these commissures, especially the anterior and posterior commissures, is commonly used in image fusion for deep brain stimulation surgery or radiosurgery. Currently, they are hardly implicated in resective surgery; however,

http://dx.doi.org/10.5772/67897

in future, they may be appropriate white matter targets for brain stimulation to modulate functions arising from certain part of both hemispheres.

5. Concept of death related to brainwaves

Once knowing the origin for the brainwaves (deep brain area), cortical functions and its fast hemispheric transfer of information (superficial brain area), perhaps then, the concept of death would easily be understood. If someone cut off his 'leg or hand or mouth or face, he shall not die', but if someone injured the core or deep area of the brain (the seat of soul area), or the cardiopulmonary system, death is likely. Therefore, death seems to be associated with two main human organs—the brain and the heart. Based on this, there are two types of deaths: (a) cardiac or circulatory death and (b) brain death. It seems that in both types of deaths, the anatomical region that concerns the brainwave origin or the greater limbic system is notably involved [66–68].

Brain death is associated with cessation of all brain functions. All points related to brain death are essentially documenting dysfunction in the greater limbic system (or the seat of soul area), such as: (a) conscious level, (b) autonomic disturbances, (c) absent brainstem reflexes, (d) flattened cortical brainwaves (bihemispheric dysfunction) and (e) disturbance in vital signs (noteworthy that these vital signs such as respiration, heart rate and blood pressure can be preserved by ventilatory support and medications in brain death). A dysfunction in anatomical region that controls brainwave rhythm would finally cause flattened cortical brainwaves. This may indirectly signify that cortical brainwaves have originated from deep structures inside the brain (the greater limbic system), and brain functions have indeed originated deep within the centre of the brain, involving anatomical areas that have vast networks with the cortices. On the other hand, for cardiac death, the cardiopulmonary system stops functioning and hence after few minutes (3–5 min), the brain also starts to stop functioning. This type of death is what most lay people think of when they think about the definition of death. Therefore, in documenting cardiac death, the person's pupils are commonly noted as fixed and dilated, and the vital signs (wavy items such as heart rate, blood pressure, respiration) are absent. Therefore, what seems initially as cardiac death is in fact related to the death of the brain too. All these indirectly denote that the brain is superior than the heart, and the seat (centre) of human soul likely resides in the brain at the greater limbic area; it may not be the observable anatomical structures in this area per se but instead is an 'unseen' element at this particular deep-centred anatomical area (noteworthy that the initial historical discussions on humans' seat of the soul and the greater limbic system are mainly meant for death status and unique human behaviour). In conclusion, five points are worth being emphasized and they are: (a) the brain seems superior than the heart because of the following reasons: (i) the status of the brain function is the most important in determining death of a person, (ii) vital signs of the cardiopulmonary system such as heart rate, blood pressure and respiration (wavy items) can be supported by a machine and medications, (iii) in contrast to point (ii) above, the flattened brainwaves seem unlikely reversible to wavy brainwaves in a dead person, and, perhaps, no machine might be able to cause reappearance of 'persistent wavy' brainwaves in a dead person, (b) waves (ups and downs, downs and ups, right-left, left-right oscillations) may be 'indirect' manifestations of the soul; once dead, all waves are flattened and finally all atoms stop oscillating (non-wavy), and physical dimension starts to disappear. Remember that atoms can behave either as particles or waves. The phenomenon is known as wave-particle duality for an atom [69, 70], (c) brainwaves can be regarded as a way to 'visualize thought' as 'images'; therefore, more studies are needed to correlate brainwaves with brain anatomy, and, indeed, advanced technology is obviously needed to enable scientists examining the deep brainwaves non-invasively and correlating them with cortical (superficial) brainwaves, brain anatomy and functions, (d) all are waves (ups and downs, energy, life, the will to live, an indirect manifestation of soul or all is the soul) and finally (e) studies on waves, oscillations, frequency and physiology (even anatomy, simply because atoms can also behave as waves) could in fact be studies related to the soul.

6. Conclusions

This chapter stresses that the central nervous system could indeed lie in the microgravity environment. The importance of this notion includes studying the brain, brain cells or tissues or, specifically, the neural stem cells in a buoyant environment. The microgravity environment of CNS has also caused bending to occur at the mid-brain level involving a set of deep anatomical structures that lie 'close to the ventricles' and link to various brain functions, including control of consciousness and memory, and even are related to death. Noteworthy that this deep brain area also seems to regulate cortical brainwave rhythms and has close connectivity with two brain hemispheres. This bilateral hemispheric connectivity was studied on 13 clinical patients' brainwaves. Bilateral hemispheric brainwave responses were observed in tasks that were related to cognition for language, sensorimotor and auditory-visual functions. Topographical or brain lobe MEG wave representations and Matlab-SPM and BESA-based brainwave spectral analysis revealed that each task has a hemispheric specialization or lateralization, which suggests that there is fast brainwave information transfer between the two brain hemispheres via the commissural system as well as an efficient information integration system in each brain hemisphere. Therefore, one may view that cortical brain functions could have originated deep and within the centre of the brain. With advancement in neurotechnology, we hope that our hypotheses, clinical findings and conclusions drawn from this chapter may form the basis to study further the deep anatomical brain structures in relation to brain functions, neurocognitions and the seat of human soul.

Acknowledgements

The content of this chapter is partially obtained from study data using short-term grant (ref: 304/PPSP/61312142 from Universiti Sains Malaysia (USM)) and is approved by Human Research Ethics Committee, School of Medical Sciences, USM, Kubang Kerian, Kelantan, Malaysia.

http://dx.doi.org/10.5772/67897

Author details

Zamzuri Idris[1,2]*, Faruque Reza[2] and Jafri Malin Abdullah[1,2]

*Address all correspondence to: neuroscienceszamzuri@yahoo.com

1 Center for Neuroscience Service and Research, School of Medical Sciences, Universiti Sains Malaysia, Kelantan, Malaysia

2 Department of Neurosciences, School of Medical Sciences, Universiti Sains Malaysia, Kelantan, Malaysia

References

[1] Hartmann P, Ramseier A, Gudat F, Mihatsch MJ, Polasek W. Normal weight of the brain in adults in relation to age, sex, body height and weight. Der Pathologe. 1994;**15**(3):165-170.

[2] Noback CNL, Robert JD, David AR. The human nervous system: structure and function. Humana Press, New Jersey: ISBN-13: 978-1588290403.

[3] Idris Z, Mustapha M, Abdullah JM. Microgravity environment and compensatory: decompensatory phases for intracranial hypertension form new perspectives to explain mechanism underlying communicating hydrocephalus and its related disorders. Asian J Neurosurg. 2014;**9**(1):7-13.

[4] Idris Z, Muzaimi M, Ghani RI, Idris B, Kandasamy R, Abdullah JM. Principles, anatomical origin and applications of brainwaves: a review, our experience and hypothesis related to microgravity and the question on soul. J Biomed Sci Eng. 2014;**7**(8):11.

[5] Idris Z. Searching for the origin through central nervous system: areview and thought which related to microgravity, evolution, big bang theory and universes, soul and brainwaves, greater limbic system and seat of the soul. Malays J Med Sci. 2014;**21**(4):4-11.

[6] Liu PC, Liu K, Liu JF, Xia K, Chen LY, Wu X. Transfection of the IHH gene into rabbit BMSCs in a simulated microgravity environment promotes chondrogenic differentiation and inhibits cartilage aging. Oncotarget. 2016;**7**(39):62873-85.

[7] Meigal AY. Synergistic action of gravity and temperature on the motor system within the lifespan: a "baby astronaut" hypothesis. Med Hypotheses. 2013;**80**(3):275-283.

[8] Chen J, Liu R, Yang Y, Li J, Zhang X, Wang Z, et al. The simulated microgravity enhances the differentiation of mesenchymal stem cells into neurons. Neurosci Lett. 2011;**505**(2):171-175.

[9] Coghill RC, Gilron I, Iadarola MJ. Hemispheric lateralization of somatosensory processing. J Neurophysiol. 2001;**85**(6):2602-2612.

[10] Gandour J, Tong Y, Wong D, Talavage T, Dzemidzic M, Xu Y, et al. Hemispheric roles in the perception of speech prosody. NeuroImage. 2004;**23**(1):344-357.

[11] Guidotti R, Del Gratta C, Baldassarre A, Romani GL, Corbetta M. Visual learning induces changes in resting-state fMRI multivariate pattern of information. J Neurosci. 2015;**35**(27): 9786-9798.

[12] Mehta JP, Verber MD, Wieser JA, Schmit BD, Schindler-Ivens SM. The effect of movement rate and complexity on functional magnetic resonance signal change during pedaling. Motor Control. 2012;**16**(2):158-175.

[13] Mohades SG, Struys E, Van Schuerbeek P, Baeken C, Van De Craen P, Luypaert R. Age of second language acquisition affects nonverbal conflict processing in children: an fMRI study. Brain Behav. 2014;**4**(5):626-642.

[14] Zhang Y, Wang T, Huang P, Li D, Qiu J, Shen T, et al. Free language selection in the bilingual brain: an event-related fMRI study. Sci Rep. 2015;**5**:11704.

[15] Santoro G, Wood MD, Merlo L, Anastasi GP, Tomasello F, Germano A. The anatomic location of the soul from the heart, through the brain, to the whole body, and beyond: a journey through Western history, science, and philosophy. Neurosurgery. 2009;**65**(4):633-643; discussion 43.

[16] Dolan B. Soul searching: a brief history of the mind/body debate in the neurosciences. Neurosurg Focus. 2007;**23**(1):E2.

[17] Nieuwenhuys R, Veening JG, van Domburg P. Core and paracores; some new chemoarchitectural entities in the mammalian neuraxis. Acta Morphol Neerlando-Scand. 1988;**26**(2-3):131-163.

[18] Nieuwenhuys R, Voogd J, Van Huijzen C. Greater limbic system. In The human central nervous system (4th ed.). Wurzburg, Berlin and Heidelberg: Springer-Verlag; 2008. pp. 917-946.

[19] Gadde J, Dross P, Spina M. Syndrome of the trephined (sinking skin flap syndrome) with and without paradoxical herniation: a series of case reports and review. Delaware Med J. 2012;**84**(7):213-218.

[20] Mohammadi A, Ahmadian A, Azar AD, Sheykh AD, Amiri F, Alirezaie J. Estimation of intraoperative brain shift by combination of stereovision and doppler ultrasound: phantom and animal model study. Int J Comput Assist Radiol Surg. 2015;**10**(11):1753-1764.

[21] Ji S, Fan X, Roberts DW, Hartov A, Paulsen KD. Cortical surface shift estimation using stereovision and optical flow motion tracking via projection image registration. Med Image Anal. 2014;**18**(7):1169-1183.

[22] Wang N, Wang H, Chen J, Zhang X, Xie J, Li Z, et al. The simulated microgravity enhances multipotential differentiation capacity of bone marrow mesenchymal stem cells. Cytotechnology. 2014;**66**(1):119-131.

[23] Magoun HW. An ascending reticular activating system in the brain stem. AMA Arch Neurol Psychol. 1952;**67**(2):145-154; discussion 67-71.

[24] Leon Dănăilă and Mihail Lucian Pascu (2013). Contributions to the Understanding of the Neural Bases of the Consciousness, Clinical Management and Evolving Novel Therapeutic Strategies for Patients with Brain Tumors, Dr. Terry Lichtor (Ed.), Rijeka Croatia: InTech, doi: 10.5772/52688. Available from: http://www.intechopen.com/books/clinical-management-and-evolving-novel-therapeutic-strategies-for-patients-with-brain-tumors/contributions-to-the-understanding-of-the-neural-bases-of-the-consciousness

[25] Englot DJ, Hassnain KH, Rolston JD, Harward SC, Sinha SR, Haglund MM. Quality of-life metrics with vagus nerve stimulation for epilepsy from provider survey data. Epilepsy Behav. 2016;**66**:4-9.

[26] Hulsey DR, Riley JR, Loerwald KW, Rennaker RL, Kilgard MP, Hays SA. Parametric characterization of neural activity in the locus coeruleus in response to vagus nerve stimulation. Exp Neurol. 2016;**289**:21-30.

[27] Bartolomei F, Bonini F, Vidal E, Trebuchon A, Lagarde S, Lambert I, et al. How does vagal nerve stimulation (VNS) change EEG brain functional connectivity? Epilepsy Res. 2016;**126**:141-146.

[28] Idris Z, Kandasamy R, Reza F, Abdullah JM. Neural oscillation, network, eloquent cortex and epileptogenic zone revealed by magnetoencephalography and awake craniotomy. Asian J Neurosurg. 2014;**9**(3):144-152.

[29] Allen P, Modinos G, Hubl D, Shields G, Cachia A, Jardri R, et al. Neuroimaging auditory hallucinations in schizophrenia: from neuroanatomy to neurochemistry and beyond. Schizophrenia Bull. 2012;**38**(4):695-703.

[30] Bingel U, Quante M, Knab R, Bromm B, Weiller C, Buchel C. Single trial fMRI reveals significant contralateral bias in responses to laser pain within thalamus and somatosensory cortices. NeuroImage. 2003;**18**(3):740-748.

[31] Shergill SS, Brammer MJ, Williams SC, Murray RM, McGuire PK. Mapping auditory hallucinations in schizophrenia using functional magnetic resonance imaging. Arch Gen Psychol. 2000;**57**(11):1033-1038.

[32] Silbersweig DA, Stern E, Frith C, Cahill C, Holmes A, Grootoonk S, et al. A functional neuroanatomy of hallucinations in schizophrenia. Nature. 1995;**378**(6553):176-179.

[33] Okada YC, Wu J, Kyuhou S. Genesis of MEG signals in a mammalian CNS structure. Electroencephalogr Clin Neurophysiol. 1997;**103**(4):474-485.

[34] Dronkers NF, Plaisant O, Iba-Zizen MT, Cabanis EA. Paul Broca's historic cases: high resolution MR imaging of the brains of Leborgne and Lelong. Brain. 2007;**130**(Pt 5):1432-1441.

[35] Herve PY, Zago L, Petit L, Mazoyer B, Tzourio-Mazoyer N. Revisiting human hemispheric specialization with neuroimaging. Trends Cogn Sci. 2013;**17**(2):69-80.

[36] Buchel C, Raedler T, Sommer M, Sach M, Weiller C, Koch MA. White matter asymmetry in the human brain: a diffusion tensor MRI study. Cereb Cortex. 2004;**14**(9):945-951.

[37] James JS, Kumari SR, Sreedharan RM, Thomas B, Radhkrishnan A, Kesavadas C. Analyzing functional, structural, and anatomical correlation of hemispheric language lateralization in healthy subjects using functional MRI, diffusion tensor imaging, and voxel-based morphometry. Neurol India. 2015;**63**(1):49-57.

[38] Takao H, Abe O, Yamasue H, Aoki S, Sasaki H, Kasai K, et al. Gray and white matter asymmetries in healthy individuals aged 21-29 years: a voxel-based morphometry and diffusion tensor imaging study. Human Brain Map. 2011;**32**(10):1762-1773.

[39] Vernooij MW, Smits M, Wielopolski PA, Houston GC, Krestin GP, van der Lugt A. Fiber density asymmetry of the arcuate fasciculus in relation to functional hemispheric language lateralization in both right- and left-handed healthy subjects: a combined fMRI and DTI study. NeuroImage. 2007;**35**(3):1064-1076.

[40] Fiori S, Staudt M, Pannek K, Borghetti D, Biagi L, Scelfo D, et al. Is one motor cortex enough for two hands? Dev Med Child Neurol. 2015. Oct;**57**(10):977-80. doi: 10.1111/dmcn.12817. Epub 2015 Jun 22.

[41] Geven LI, de Kleine E, Willemsen AT, van Dijk P. Asymmetry in primary auditory cortex activity in tinnitus patients and controls. Neuroscience. 2014;**256**:117-125.

[42] Kam JW, Nagamatsu LS, Handy TC. Visual asymmetry revisited: mind wandering preferentially disrupts processing in the left visual field. Brain Cogn. 2014;**92C**:32-38.

[43] Kim YH, Jang SH, Han BS, Kwon YH, You SH, Byun WM, et al. Ipsilateral motor pathway confirmed by diffusion tensor tractography in a patient with schizencephaly. Neuroreport. 2004;**15**(12):1899-1902.

[44] Mutha PK, Haaland KY, Sainburg RL. Rethinking motor lateralization: specialized but complementary mechanisms for motor control of each arm. PLoS One. 2013;**8**(3):e58582.

[45] Rangarajan V, Hermes D, Foster BL, Weiner KS, Jacques C, Grill-Spector K, et al. Electrical stimulation of the left and right human fusiform gyrus causes different effects in conscious face perception. J Neurosci. 2014;**34**(38):12828-12836.

[46] Asanowicz D, Smigasiewicz K, Verleger R. Differences between visual hemifields in identifying rapidly presented target stimuli: letters and digits, faces, and shapes. Front Psychol. 2013;**4**:452.

[47] Hori T, Kawamata T, Amano K, Aihara Y, Ono M, Miki N. Anterior interhemispheric approach for 100 tumors in and around the anterior third ventricle. Neurosurgery. 2010;**66**(3 Suppl Operative):65-74.

[48] Verleger R, Sprenger A, Gebauer S, Fritzmannova M, Friedrich M, Kraft S, et al. On why left events are the right ones: neural mechanisms underlying the left-hemifield advantage in rapid serial visual presentation. J Cogn Neurosci. 2009;**21**(3):474-488.

[49] Jung-Beeman M. Bilateral brain processes for comprehending natural language. Trends Cogn Sci. 2005;**9**(11):512-518.

[50] Fabri M, Pierpaoli C, Barbaresi P, Polonara G. Functional topography of the corpus callosum investigated by DTI and fMRI. World J Radiol. 2014;**6**(12):895-906.

[51] Fabri M,Polonara G. Functional topography of human corpus callosum: an FMRI mapping study. Neural Plast. 2013;**2013**:251308.

[52] Rakic P,Yakovlev PI. Development of the corpus callosum and cavum septi in man. J Comp Neurol. 1968;**132**(1):45-72.

[53] Raybaud C. The corpus callosum, the other great forebrain commissures, and the septum pellucidum: anatomy, development, and malformation. Neuroradiology. 2010;**52**(6):447-477.

[54] Suarez R, Gobius I, Richards LJ. Evolution and development of interhemispheric connections in the vertebrate forebrain. Front Human Neurosci. 2014;**8**:497.

[55] Deoni SC, Mercure E, Blasi A, Gasston D, Thomson A, Johnson M, et al. Mapping infant brain myelination with magnetic resonance imaging. J Neurosci. 2011;**31**(2):784-791.

[56] Provenzale JM, Isaacson J, Chen S. Progression of corpus callosum diffusion-tensor imaging values during a period of signal changes consistent with myelination. Am J Roentgenol. 2012;**198**(6):1403-1408.

[57] Rauch RA,Jinkins JR. Analysis of cross-sectional area measurements of the corpus callosum adjusted for brain size in male and female subjects from childhood to adulthood. Behav Brain Res. 1994;**64**(1-2):65-78.

[58] Barkovich AJ,Kjos BO. Normal postnatal development of the corpus callosum as demonstrated by MR imaging. Am J Neuroradiol. 1988;**9**(3):487-491.

[59] Genc E, Ocklenburg S, Singer W, Gunturkun O. Abnormal interhemispheric motor interactions in patients with callosal agenesis. Behav Brain Res. 2015. Oct 15;**293**:1-9. doi: 10. 1016/j.bbr.2015.07.016. Epub 2015 Jul 14.

[60] Hasegawa T, Yamada K, Morimoto M, Morioka S, Tozawa T, Isoda K, et al. Development of corpus callosum in preterm infants is affected by the prematurity: in vivo assessment of diffusion tensor imaging at term-equivalent age. Pediatr Res. 2011;**69**(3):249-254.

[61] Mathew P, Pannek K, Snow P, D'Acunto MG, Guzzetta A, Rose SE, et al. Maturation of corpus callosum anterior midbody is associated with neonatal motor function in eight preterm-born infants. Neural Plast. 2013;**2013**:359532.

[62] Sztriha L. Spectrum of corpus callosum agenesis. Pediatr Neurol. 2005;**32**(2):94-101.

[63] Shiramizu H, Hori T, Matsuo S, Niimura K, Yoshimoto H, Ishida A, et al. Anterior callosal section is useful for the removal of large tumors invading the dorsal part of the anterior third ventricle: operative technique and results. Neurosurg Rev. 2013;**36**(3):467-475.

[64] Jenssen S, Sperling MR, Tracy JI, Nei M, Joyce L, David G, et al. Corpus callosotomy in refractory idiopathic generalized epilepsy. Seizure. 2006;**15**(8):621-629.

[65] Kasasbeh AS, Smyth MD, Steger-May K, Jalilian L, Bertrand M, Limbrick DD. Outcomes after anterior or complete corpus callosotomy in children. Neurosurgery. 2014;**74**(1):17-28; discussion.

[66] Goila AK, Pawar M. The diagnosis of brain death. Indian J Crit Care Med. 2009;**13**(1):7-11.

[67] De Georgia MA. History of brain death as death: 1968 to the present. J Crit Care. 2014; **29**(4):673-678.

[68] Fishman GI, Chugh SS, Dimarco JP, Albert CM, Anderson ME, Bonow RO, et al. Sudden cardiac death prediction and prevention: report from a National Heart, Lung, and Blood Institute and Heart Rhythm Society Workshop. Circulation. 2010;**122**(22):2335-2348.

[69] Arndt M, Nairz O, Vos-Andreae J, Keller C, van der Zouw G, Zeilinger A. Wave-particle duality of C(60) molecules. Nature. 1999;**401**(6754):680-682.

[70] Gerlich S, Eibenberger S, Tomandl M, Nimmrichter S, Hornberger K, Fagan PJ, et al. Quantum interference of large organic molecules. Nat Commun. 2011;**2**:263.

Chapter 5

Anatomy of Extramuscular Soleus Veins: Clinical Impact

José Aderval Aragão, Francisco Prado Reis,
Felipe Matheus Sant'Anna Aragão,
Iapunira Catarina Sant'Anna Aragão and
Sydney Correia Leão

Additional information is available at the end of the chapter

http://dx.doi.org/10.5772/intechopen.68824

Abstract

The venous system of the lower limbs has great structural and functional anatomical complexity which must be considered in different dysfunctions of this system. This complexity lies mainly in the venous return, which is changed from the upright position and ambulation and other factors such as level of physical activity, heart function, circulating blood volume, and ambient temperature. Anatomical description of soleus veins (SV) has received little attention from books' anatomy texts. These veins are intramuscular deep veins and known as the main chamber of the calf pump. Soleus veins have been implicated as the site for deep vein thrombosis (DVT). Detailed anatomical knowledge is required for early diagnosis using noninvasive ultrasound techniques. In the present work, we describe the anatomy of the veins that emerge from the ventral surface of the soleus muscle. Twenty-eight soleus muscles were dissected and 543 veins were found. The number of veins per leg ranged from 7 to 38. The distribution of these veins per quadrant ranged from 0 to 12. The greatest number of veins occurred in the upper lateral quadrant. Most of the soleus veins drained into the posterior tibial and fibular veins. The mean length of the soleus veins ranged from 0.907 to 2.804 cm. We conclude that there is a wide variability in the distribution of soleus veins through the soleus muscle and its quadrants. The majority of the soleus veins drain into the tibial and fibular veins.

Keywords: anatomy, veins, soleus muscle

1. Introduction

The venous system of the lower limbs has biological characteristics and structural and functional complexity that can only be considered when addressing situations that lead to their dysfunction. The anatomy and physiology of the venous system of lower limbs are complex and its main characteristic is the fact that the venous return is influenced by change position, standing position, and ambulation. Venous return also depends on other factors, such as physical activity, cardiac function, circulating blood volume, and ambient temperature. These biological characteristics, belonging solely to human species, are important structural and functional complexity, since when man has adopted bipedal position as a preferred mode of locomotion, as does not possess a suitable elastic and fibrous tissue system in the lower limbs, adapted from efficient manner for the requirements of this position. The anatomical description of the soleus veins (SV) has received little attention from texts of anatomy books. Paturet [1] described the SV as satellites of the arteries. These veins, like the gastrocnemius veins, are studied by the anatomical and functional point of view as intramuscular deep vein, being known as the main chamber of the calf pump [2].

Kwakye [3] and Van Limborgh and Kwakye [4] divided the veins that receive drainage of the soleus muscle into two main groups: the posterior tibial and fibular. The fibular group receive two larger longitudinal veins, a lateral and another intermediate, that would have originated in the lower half of the soleus muscle across several small veins and end up in the posterior tibial veins.

For Kobak and Lev [5], the veins of the soleus muscle would drain mainly to the fibular veins as small soleus veins would end in the posterior tibial and fibular veins. Cocket [6] and Dodd and Cocket [7] described the venous sinuses formed inside the soleus muscle that would drain for short and loose veins that finally outfall in the tibial and fibular veins later.

According to Ukhov [8], intramuscular SV form three collector trunks: medial, intermediate, and lateral. In these collectors, the medial would be predominant, followed by intermediate and the lateral. At the height of the lower third of the soleus muscle, two to four veins that would pass by the posterior tibial veins emerge; rarely, it would happen with the presence of one venous trunk or four more veins.

The presence of venous sinuses inside the soleus muscle, ending up in venous collector trunks, was described by several authors [3, 4, 6–9]. Abramova and Chilaia [10] reported that more often there would be three venous sinuses in the soleus muscle: a lateral; midline, which would end the fibular vein; and medial, which would end in the posterior tibial veins.

Through a phlebographic study, Sequeira et al. [11] reported an average of 11.72 SV on the right leg and 10.68 on the left leg. In dissecting bodies, Sequeira et al. [12] reported an average of 46.8 per leg. They studied the veins emerging from the anterior surface of the soleus muscle.

White et al. [13] reported that the calf muscle pump has been frequently discussed but incompletely defined. Functionally, it represents the mechanism by which blood in the deep calf veins is propelled cephalad. Physiologically, compartment pressures do not rise sufficiently during ambulation to adequately compress the deep calf veins and displace the blood they contain.

Black [14] reported that the deep veins of the calf include the tibial, peroneal, soleal, and gastrocnemius veins. The anterior tibial, posterior tibial, and peroneal veins are generally paired and are located on either side of a corresponding artery. Venous sinusoids within the deep calf musculature coalesce to form the soleal and gastrocnemius venous plexuses. These muscular venous sinuses are the primary collecting system of the calf muscle pump. Soleal sinuses typically communicate with the posterior tibial veins, whereas the gastrocnemius network coalesces into paired gastrocnemius veins that drain directly into the popliteal. The anterior and posterior tibial veins join with the peroneal veins to become the popliteal vein.

For Henry and Satiani [15], the calf muscle veins are deep veins in the distal lower extremity that are nonpaired and not associated with named tibial arteries. These veins make up a complex venous system of the musculature of the posterior leg and include the soleal and gastrocnemius veins that run as sinusoids within the muscles of the same name. The soleal sinusoids may drain into the mid peroneal or posterior tibial veins, whereas the gastrocnemius sinusoids may empty directly into the popliteal vein. In addition, these veins may communicate with the short saphenous veins through a series of perforators.

Calf muscle pump, according to Recek [16], is the motive force enhancing the return of the venous blood from the lower extremity to the heart. It causes displacement of the venous blood in both vertical and horizontal directions and generates ambulatory pressure gradient between thigh and lower leg veins and bidirectional streaming within calf perforators.

According to Uhl and Gillot [17], the muscular pumps are the true peripheral heart of the venous system of the lower limbs and play a crucial role in the venous return. The basic function of the venous system of the lower limbs is to ensure the return of the blood from the peripheral tissues to the heart and that in order to be efficient, the venous system is based on two mechanisms: the normal functioning of the venous valves (anti-reflux system against gravity) and a complex system of impulse-aspiration pumps, so-called venomuscular pumps.

Uhl and Gillot [17], highlighting the description 30 years ago by Gardner and Fox [18], reported that these pumps can be divided into four main parts, creating together a true chain of synchronized events: the foot pump, located in the lateral plantar veins; the leg pump, located in the soleus muscle; the gastrocnemial pump, acting at the popliteal level above the knee, these two latter pumps together are the calf pump, the most important pump of the limb; and finally, the thigh pumps: semimembranosus, biceps (posteriorly), and quadriceps muscle (anteriorly). The synchronization of the different venomuscular pumps during walk is crucial: the foot, then leg, popliteal, and finally thigh pumps.

Many authors have paid attention to the SV veins with respect to its role in the investigation of the location of deep vein thrombosis in the calf [12, 19–25]. Thus, it has been a very usual anatomical study of the venous drainage of the soleus muscle from the knowledge of intramuscular veins.

2. Material and methods

Twenty-eight legs from 14 adult male human cadavers were used, which had been fixed and preserved in 10% formalin solution. The material was used in conformity with Law 8501 of November 30, 1992, which provides for the utilization of unclaimed cadavers for scientific research or study purposes. The present study was approved by the Research Ethics Committee of the Health Sciences University of Alagoas, Brazil, under protocol no. 038/02. Cadavers that presented macroscopically detectable pathological alterations on the lower limbs were excluded from the study.

The anatomical layers of the posterior region of the leg were dissected until the gastrocnemius muscle and the posterior surface of the soleus muscle had been exposed. On this posterior face, two transversal lines were traced out to divide the muscle into three levels: upper, middle, and lower. A median longitudinal line was traced out to intersect with the transversal lines, thereby resulting in the division of the surface of the soleus muscle into six quadrants: superior medial (QSM) and superior lateral (QSL); middle medial (QMM) and middle lateral (QML); and inferior medial (QIM) and inferior lateral (QIL), as shown in **Figure 1**. The muscle was taken out distally, and the veins without the aid of optical instruments were dissected on their ventral surface in the distal-to-proximal direction. A digital pachymeter was utilized to measure the length of all the dissected veins. The anatomical findings were recorded in tables and by means of digital photographs.

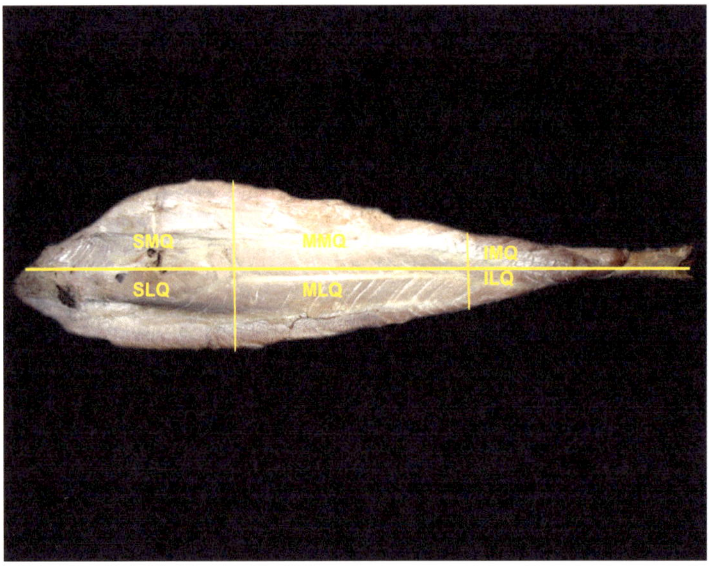

Figure 1. The soleus muscle divided into six quadrants. SMQ, superior medial quadrant; SLQ, superior lateral quadrant; MMQ, middle medial quadrant; MLQ, middle lateral quadrant; IMQ, inferior medial quadrant; and ILQ, inferior lateral quadrant.

3. Results

These results corresponded to the dissection of 28 legs. We found a total of 543 SV that emerged from the anterior face of the soleus muscle: 268 in the right legs and 275 in the left legs. The mean numbers of veins in the right and left legs were similar. In the right leg, the numbers ranged from 8 to 38, with a mean of 19.1. In the left leg, the numbers ranged from 7 to 36, with a mean of 19.6 (**Table 1**).

The distribution of SV by level and quadrant presented variations, with a maximum range of 12. The variations were less accentuated between the quadrants of the lower level of the muscle and greater between the middle quadrants and superior quadrants. The smallest number

Leg	Number of veins	Range	Mean
Left	268	8–38	19.142
Right	275	7–36	19.642
Total	543		

Table 1. Veins that emerge from the ventral face of the soleus muscle.

of SV per leg and quadrant (16) was found in the QIL of the right leg and the larger (77) in the QMM of the right leg. In these same quadrants, the lowest (1.1) and highest (5.5) means (**Table 2**) were found.

The SV most frequently drained into the medial posterior tibial vein (VTPM, **Figure 2**), lateral posterior tibial vein (VTPL, **Figure 3**), lateral fibular vein (VFL, **Figure 4**), and medial fibular vein (VFM, **Figure 5**). On the other hand, the short saphenous vein (VSP) and the tibiofibular trunk (TTF) received the smallest numbers of SV. Some SV also terminated simultaneously in more than one vein (**Figure 6**). This finding was most frequent in relation to the VTPM and VTPL (**Table 3**). Varying numbers of SV also terminated in the medial and lateral anterior tibial veins (VTAM, VTAL), SV, gastrocnemius vein (VG **Figure 7**), and popliteal vein (VP **Figure 8**). In 43 soleus veins, it was not possible to recognize their termination.

Quadrant	Leg	Number of veins	Variation	Mean
IMQ	Right	18	0–4	1.285
	Left	28	0–4	2.000
Total		**46**		
ILQ	Right	16	0–4	1.142
	Left	23	0–4	1.642
Total		**39**		
MMQ	Right	77	2–12	5.500
	Left	57	0–8	4.071
Total		**134**		
MLQ	Right	31	0–6	2.214
	Left	37	0–8	2.642
Total		**68**		
SMQ	Right	56	1–12	4.000
	Left	54	0–8	3.857
Total		**110**		
SLQ	Right	70	0–8	5.000
	Left	76	0–11	5.428
Total		**146**		

Table 2. Number, variation, and mean number of soleus veins per quadrant and legs.

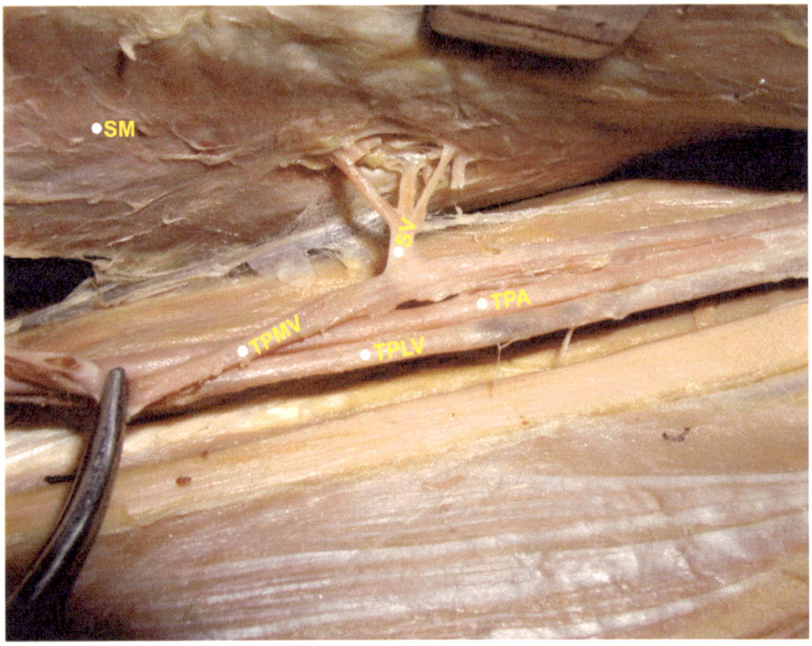

Figure 2. Soleus vein termination into the medial posterior tibial vein. SM, soleus muscle; SV, soleus vein; TPMV, tibial posterior medial vein; TPLV, tibial posterior lateral vein; and TPA, tibial posterior artery.

Figure 3. Soleus vein termination into the lateral posterior tibial vein. SM, soleus muscle; SV, soleus vein; TPMV, tibial posterior medial vein; TPLV, tibial posterior lateral vein; TPA, tibial posterior artery; and TN, tibial nerve.

http://dx.doi.org/10.5772/intechopen.68824

Figure 4. Soleus vein termination into the lateral fibular vein. SM, soleus muscle; SV, soleus vein; FLV, fibular lateral vein; and FA, fibular artery.

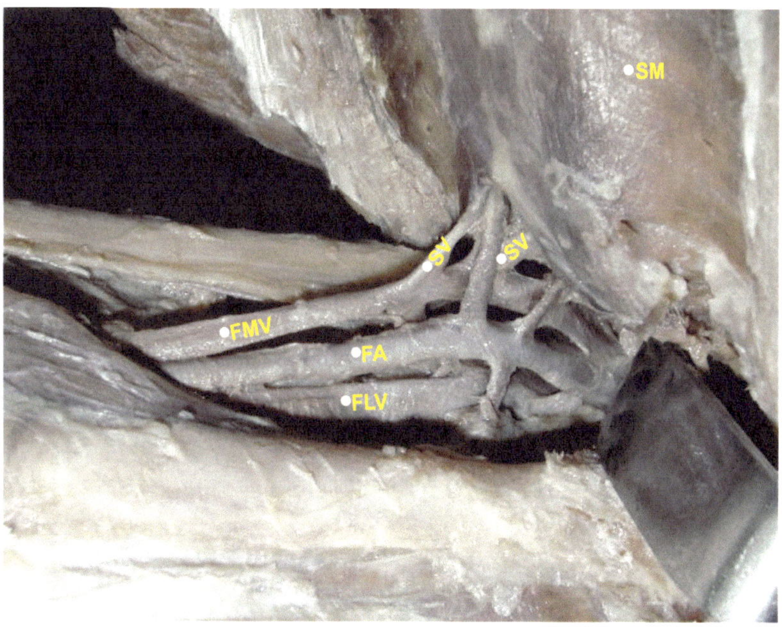

Figure 5. Soleus vein termination into the medial fibular vein. SM, soleus muscle; SV, soleus vein; FMV, fibular medial vein; FLV, fibular lateral vein; and FA, fibular artery.

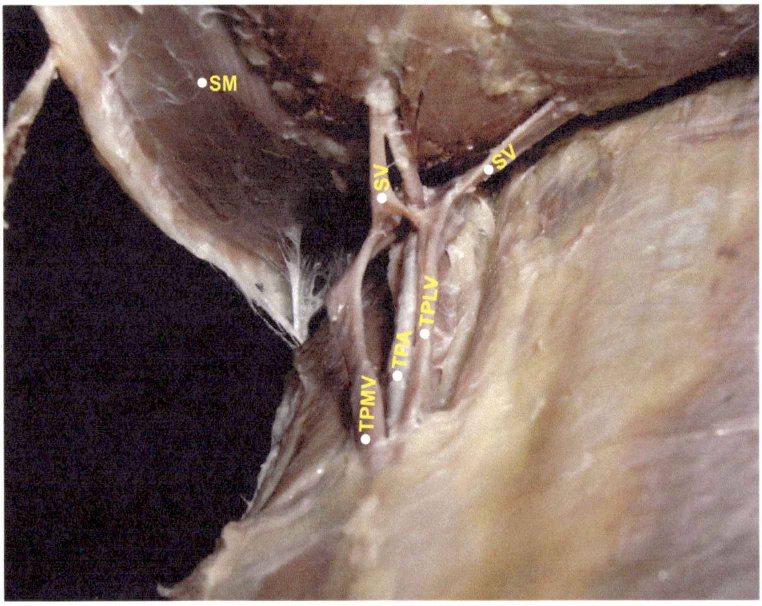

Figure 6. Simultaneous termination of the soleus vein into the medial and lateral posterior tibial veins. SM, soleus muscle; SV, soleus vein; TPMV, tibial posterior medial vein; TPLV, tibial posterior lateral vein; and TPA, tibial posterior artery.

Veins	Right leg (n = 14)	Left leg (n = 14)	Total
TPMV	75	66	**141**
TPLV	55	51	**106**
FMV	20	24	**44**
FLV	58	45	**103**
TAMV	09	02	**11**
TALV	02	03	**05**
PV	01	04	**05**
SPV	01	01	**02**
TPMV + TPLV	06	10	**16**
TPLV + FMV	01	02	**03**
FMV + FLV	04	03	**07**
TFT	00	02	**02**
SV	06	07	**13**
GV	14	05	**19**
Others	13	30	**43**

TPMV, tibial posterior medial vein; TPLV, tibial posterior lateral vein; FMV, fibular medial vein; FLV, fibular lateral vein; TAMV, tibial anterior medial vein; TALV, tibial anterior lateral vein; PV, popliteal vein; SPV, small saphenous vein; TFT, tibiofibular trunk; SV, soleus vein; GV, gastrocnemius vein

Table 3. Drainage of soleus muscle.

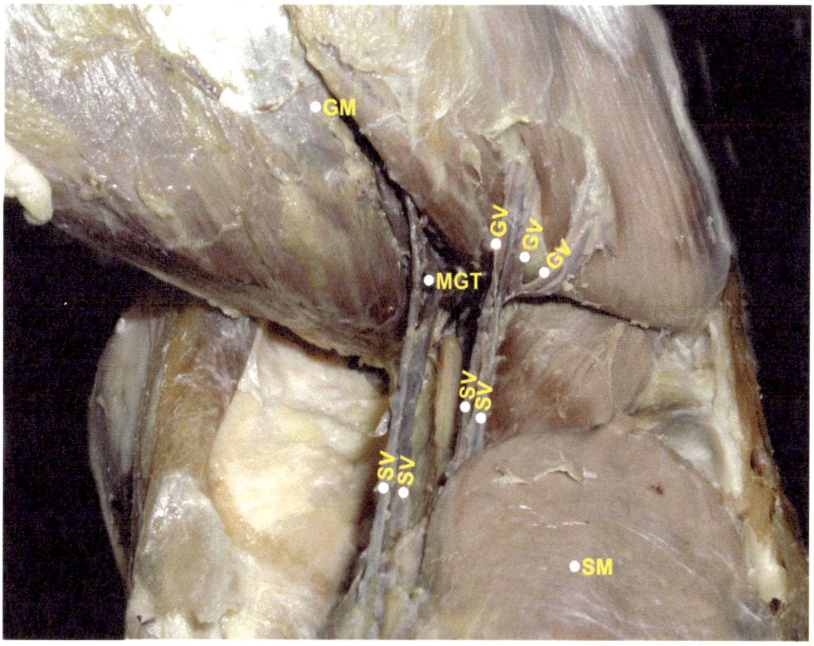

Figure 7. Soleus vein termination into the main gastrocnemius vein and trunk. SM, soleus muscle; SV, soleus vein; GM, gastrocnemius muscle; GV, gastrocnemius vein; and MGT, main gastrocnemius trunk.

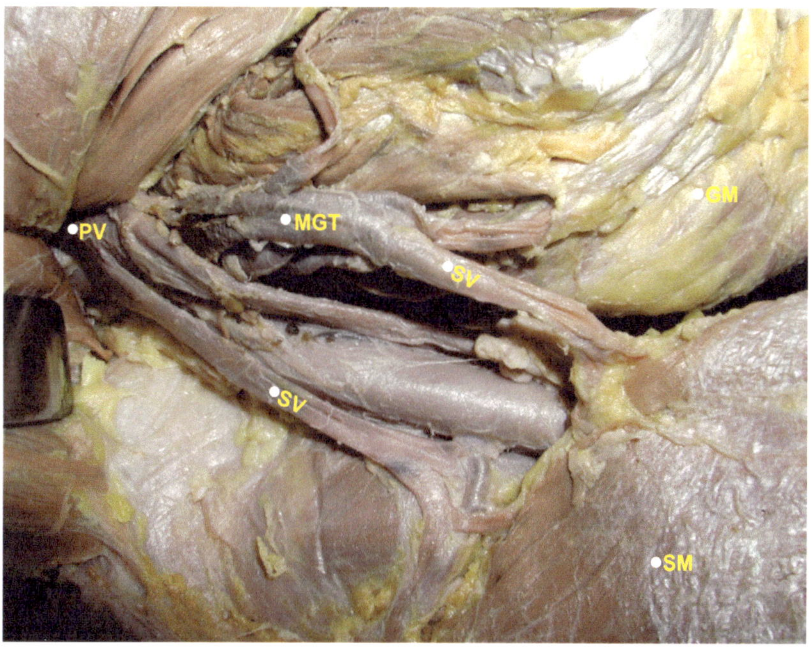

Figure 8. Soleus vein termination into the popliteal vein and main gastrocnemius trunk. SM, soleus muscle; SV, soleus vein; GM, gastrocnemius muscle; PV, popliteal vein; and MGT, main gastrocnemius trunk.

The mean lengths of the SV ranged from 0.907 to 2.804 cm. The smallest length was found in the QMM of the right leg and the greatest length in the QIL of the left leg (**Table 4**). The mean length of the SV was least (1.139 cm) in the middle level of the soleus muscle of the right leg and greatest (2.172 cm) in the lower level of the soleus muscle of the left leg (**Table 5**). This greatest mean length was also similar to what was found at the lower level of the right leg muscle.

Leg	n	QIM	QIL	QMM	QML	QSM	QSL
Right	14	1.627	2.550	0.907	1.716	1.228	1.328
Left	14	1.653	2.804	1.062	1.662	1.475	1.436
Total	28						

Table 4. Mean length (cm) of the extramuscular veins of the soleus muscle per quadrant.

Leg	n	Third		
		Inferior	Medium	Superior
Right	14	2.061	1.139	1.284
Left	14	2.172	1.339	1.454
Total	28			

Table 5. Mean length (cm) of the extramuscular veins of the soleus muscle per third of the height of the muscle.

4. Discussion

A general systematization of the SV anatomy described by (White et al. [13]), was reported by Kwakye [3]. More recently, among other authors, Uhl and Gillot [17] described an overall systematization of the anatomy of the SV. For those authors, these veins, the SV, are clearly visible inside the muscle, each part being divided into central, close to the septum and peripheral, located laterally. The medial veins of the soleus are smaller than the lateral veins of the soleus and are oriented horizontally in the peripheral part of the muscle and vertically in the central part. These vertical and central veins join the midline at the proximal part of the muscle to connect the fibular veins more laterally. The lateral view shows the large volume of the lateral veins of the soleus, directed vertically. They join in several trunks ending in the fibular veins, above the arcade of the long flexor of the hallux muscle. This explains why the fibular veins are much larger above this arcade. Below, they are contained into the fibrous, inextensible fibular canal. Above, they are dilated due to the arrival of those large lateral veins of the soleus.

Uhl and Gillot [17] concluded, in summary, that the drainage of the veins of the soleus muscle is divided into two parts: the medial veins horizontally into the posterior tibial veins and the lateral veins vertically into the fibular veins.

Uhl and Gillot [17] also reported on the finding of a, not described previously, superior vein or dorsal vein of the soleus (DVS). The specific landmark of this vein is the belly of the plantaris muscle, located between the gastrocnemius and soleus muscles. It arises from the lower part of the lateral prolongation of the linea aspera and from the oblique popliteal ligament of the knee joint. It forms a small fusiform belly, from 7 to 10 cm long, ending in a long slender tendon that crosses obliquely between the two muscles of the calf. It runs along the medial border of the calcaneus tendon to be inserted with it into the posterior part of the calcaneus.

In the present study, the anatomy of the SV was described in relation to their emergence on the anterior surface of the soleus muscle. This approach has also been performed by Sequeira et al. [11, 12, 26] and Reis et al. [27]. So in that way, most of our discussion was held at the work of these single authors, in Brazil, to carry out this type of study for soleus veins.

Kageyama et al. [28] and Ro and Kageyama [29] reported that the soleal vein contains over 10 multibranched veins in each leg and they are roughly subclassified into three groups: (1) centralis, (2) medialis, and (3) lateralis.

Ohgi and Ohgi [30] investigated the relationships between specific distributions of isolated thrombosis of the soleus vein sole thrombosis (SVT) and risk factors; in the vein classification, the soleus muscle was divided into six circulatory regions and intramuscular veins were divided into six groups—proximal, lateral, central, medial, distal medial, and distal lateral veins—based on the circulatory regions and deep veins communicated with intramuscular veins in these regions. Despite the similar approach, these authors did not, however, describe any morphometric data or the topography and distribution of the extramuscular soleus veins.

The mean numbers of SV per leg and per individual were quite different. We found a mean of 19.34 SV per leg and 38.68 per individual, while the means cited by those authors were 46.76 and 93.52, respectively.

We emphasize that the authors did not make any reference to variations in the numbers of veins of each of the legs. In relative terms, we found that the percentages of SV per leg were similar to those of Sequeira et al. [11, 12]. The number of SV increased from the lower level to the upper level. Together, the middle and upper levels presented a concentration of 84.34% of all of the SV, with 47.14% in the upper level. This may be related with the anatomy of the soleus muscle.

With regard to the distribution of SV per quadrant, our findings differ from those of Sequeira et al. [11, 12]. These authors found that 7.39% of them were located in the QMM and 6.65% in the QSL. These data are different from the phlebographic findings of Sequeira et al. [11], in which the SV locations were 29.1% in the QML and 27.4% in the QMM. Our findings were that 26.8% these veins were located in the QSL and 24.6% in the QMM. Among the quadrants, 53.4% of the veins were located medially and 46.59% laterally.

Around 75.7% of the SV drained into the tibial and fibular veins. This pattern appears to be in agreement with that described by authors as Van Limborgh and Kwakye [4], Kobak and Lev [5], Henry and Satiani [15], Black [14], and Uhl and Gillot [17]. Out of the 47.5% of the

SV that terminated in the tibial veins, 27.1% terminated in the VTPM and 20.3% in the VTPL. With regard to the fibular veins, 19.8% of the SV terminated in the VFL and 8.4% in the VFM. Using phlebography, Sequeira et al. [11] described a pattern of predominant termination of SV in the fibular vein. The mean length of the SV according to the level of the soleus muscle was greatest at the lower level of the left leg muscle. The smallest mean length was found at the middle level of the right leg muscle.

We present here a static anatomical aspect of the SV. The authors have used the anatomical aspect of soleal and gastrocnemius vein to explain clinically and physiopathologically the involvement of these vessels in the deep vein thrombosis (DVT) and other disorders that occur in the calf muscle pump.

Henry and Satiani [15] highlighted the significant impact of DVT on the worldwide population health. They divided deep thrombosis of the veins of the lower limbs into proximal or axial DVT and calf DVT. This discussion was related to calf muscle venous thrombosis (CMVT) or DVT involving isolated gastrocnemius and soleal vein thrombosis as well as in combination with proximal or axial DVT.

Keijsers et al. [31] admitted that the collapsibility of the veins and the dynamics of the venous valves that direct the blood toward the heart and shield hydrostatic pressure are believed to be the main physiological factors in the muscle pump effect. Thus they studied the dynamics of calf muscle pump function during a muscle contraction using the different model configurations that were reported in four sections which examine the course of the deep venous collapse, the effect of venous valves, the effect of hydrostatic pressure, and the importance of the superficial system, respectively. They concluded that the model developed was able to predict the increase in venous return during muscle contraction. As the proximal valves close during the relaxation phase, reflux is prevented, which without valves resulted in a loss of 53% of effective venous return. Furthermore the closing of the valves increases the perfusion in the relaxation phase. Finally, the inclusion of the superficial venous system demonstrates the role of the superficial veins in maintaining arterial inflow during muscle contraction and decreasing refilling time by 37% during relaxation.

Williams et al. [32], conducted a review whose primary objective was the evaluate the relationship between calf muscle pump function and the onset and progression of chronic vascular disease (CVD), using the available literature. The authors identified a correlation between calf muscle pump dysfunction and CVD, whose data implied calf muscle pump impairment as a clinical manifestation associated with symptomatic disease. According to available literature, they concluded that the linear relationship between the clinical manifestation of CVD with the ejection fraction (EF) is consistent, but it is not conclusive; it supports the association between calf muscle pump dysfunction and objective measures of CVD severity.

We believe that our main contribution was the description of the emerging veins of the anterior surface of the soleus muscle. Thus our focus was to identify the large number of veins, some of which next to arteries form vascular pedicle and were draining. These veins are likely to be more accessible to the image and surgical dissections. This can be a contribution to assist vascular and plastic surgeons as well as those radiologists and for future hemodynamic studies.

5. Final considerations

The present anatomical findings as the veins of the soleus muscle exemplify the complexity of the human venous system, particularly of the lower limbs, whose veins have significant role in physiology in the venous return as well as of its involvement in the origin and development of physiopathology of deep vein thrombosis and chronic vascular disease.

Acknowledgements

This chapter features content reproduced from authors' earlier publication on this topic [27].

Author details

José Aderval Aragão[1,2]*, Francisco Prado Reis[2], Felipe Matheus Sant'Anna Aragão[3], Iapunira Catarina Sant'Anna Aragão[3] and Sydney Correia Leão[4]

*Address all correspondence to: jaafelipe@infonet.com.br

1 Federal University of Sergipe (UFS), Aracaju, Brazil

2 Tiradentes University (UNIT), Aracaju, Brazil

3 University Center of Volta Redonda(UNIFOA), Brazil

4 Federal University of São Paulo (UNIFESP), Brazil

References

[1] Paturet G. Traité d'anatomie humaine. Paris: Masson; 1951. 859 p

[2] Browse NL, Burnand KG, Irvine AT, Wilson NM. Fisiologia e anatomia funcional. In: Browse NL, Burnand KG, Irvine AT, Wilson NM, editors. Doenças venosas. 2nd ed. Rio de Janeiro: Di-Livros; 2001. p. 47-59

[3] Kwakye LB. Gross anatomy of the soleus veins. Acta Morphologica Neerlando-Scandinavica. 1972;**9**(4):281-294. PMID: 5055989

[4] Van Limborgh J, Kwakye LB. Anatomie normale des veines soleaires chez lês personnes agées. Phlébologie. 1975;**28**(2):273-9

[5] Kobak M, Lev M. Anatomy of the deep venous system of the popliteal fossa and lower leg. AMA Archives of Surgery. 1954;**68**(3):350-354. PMID: 13137693

[6] Cockett FB. The pathology and treatment of venous ulcers of the leg. British Journal of Surgery. 1955;**43**(179):260-78. PMID: 13284261

[7] Dodd H, Cockett FB. Anatomy of the veins of the lower limb. In: Dodd H, Cockett FB, editors. The Pathology and Surgery of the Veins of the Lower Limb. 2nd ed. Edinburgh: Churchill Livingstone; 1976. pp. 18-24

[8] Ukhov II. Veins of the soleus and their relationship with the organ structure. Arkhiv Anatomii, Gistologii I Embriologii. 1974;**67**(7):104-110. PMID: 4455143

[9] Coito AMF. Particularidades anatômicas do sistema venoso dos membros inferiores. Gazeta Médica Portuguesa.1959;**12**(2):141-194

[10] Abramova AA, Chilaia SM. Structure of the venous sinuses of M. soleus and their role in the origin of thrombosis of the deep veins in the lower limbs. Arkhiv Anatomii, Gistologii I Embriologii. 1967;**53**(9):63-67. PMID: 5614568

[11] Sequeira CMG, Martins FG, Andrade JA, Mayall ACDG, Mayall RC. Estudo flebográfico das veias soleares. Revista de angiologia e cirurgia vascular 2005;**1**(1):15-23

[12] Sequeira CMG, Juliano Y, Novo NF, Mayall RC, Júnior FM. Veias soleares: bases anatômicas e seu papel na origem da trombose venosa profunda em membro inferior. Revista da Associação Médica Brasileira. 2007;**53**(4):305-310. DOI: 10.1590/S0104-42302007000400014

[13] White JV, Katz ML, Cisek P, Kreithen J. Venous outflow of the leg: Anatomy and physiologic mechanism of the plantar venous plexus. Journal of Vascular Surgery. 1996 Nov;**24**(5):819-24. PMID: 8918329

[14] Black CM. Anatomy and physiology of the lower-extremity deep and superficial veins. Techniques in Vascular and Interventional Radiology. 2014 Jun;**17**(2):68-73. PMID: 24840960. DOI: 10.1053/j.tvir.2014.02.002

[15] Henry JC, Satiani B. Calf muscle venous thrombosis: A review of the clinical implications and therapy. Vascular and Endovascular Surgery. 2014 Jul–Aug;**48**(5-6):396-401. PMID: 25027613. DOI: 10.1177/1538574414541704

[16] Recek C. Conception of the venous hemodynamics in the lower extremity. Angiology. 2006 Oct-Nov;**57**(5):556-563. PMID: 17067977. DOI: 10.1177/0003319706293117

[17] Uhl JF, Gillot C. Anatomy of the veno-muscular pumps of the lower limb. Phlebology. 2015 Apr;**30**(3):180-93. PMID: 24415543. DOI: 10.1177/0268355513517686

[18] Gardner AM, Fox RH. The venous pump of the human foot: preliminary report. Bristol Medico-Chirurgical Journal or Bristol Med Chir J 1983;**98**:109-112. PMID: 6616290

[19] Ohgi S, Tachibana M, Ikebuchi M, Kanaoka Y, Maeda T, Mori T. Pulmonary embolism in patients with isolated soleal vein thrombosis. Angiology. 1988;**49**(9):759-764. PMID: 9756428

[20] Krünes U, Teubner K, Knipp H, Holzapfel R. Thrombosis of the muscular calf veins—Reference to a syndrome which receives little attention. Vasa. 1988;**27**(3):172-175. PMID: 9747154

[21] Labropoulos N, Webb KM, Kang SS, Mansour MA, Filliung DR, Size GP, Buckman J, Baker WH. Patterns and distribution of isolated calf deep vein thrombosis. Journal of Vascular Surgery. 1999;**30**(5):787-791. PMID: 10550175

[22] Sprayegen S, Koenigsberg K, Haimovici H. Flebografia contrastada e imagens ultrasônicas venosas. In: Haimovici H, editor. Cirurgia vascular: princípios e técnicas. 4th ed. Rio de Janeiro: Di-Livros; 1999. pp. 1172-1190

[23] Hollerweger A, Macheiner P, Rettenbacher T, Gritzmann N. Sonographic diagnosis of thrombosis of the calf muscle veins and the risk of pulmonary embolism. Ultraschall in der Medizin. 2000;**21**(2):66-72. PMID: 10838706. DOI: 10.1055/s-2000-8929

[24] Garrido MBM. Anatomia médico-cirúrgica do sistema venoso dos membros inferiores. In: Maffei FHA, editor. Doenças vasculares periféricas. 3rd ed. São Paulo: Medisi; 2002. pp. 133-167

[25] Cohen BJ, Wood DL. Vasos sanguíneos e circulação do sangue. In: Cohen BJ, Wood DL, editors. O corpo humano na saúde e na doença. 1st ed. São Paulo: Manole; 2002. pp. 267-290

[26] Sequeira CMG, Miranda Jr F, Mayall RC. Estudo da terminação das veias soleas no homem. Jornal Vascular Brasileiro. 2003;**2**(S1):39

[27] Reis FP, Aragão JA, de Figueiredo LF, Miranda F Jr, Nunes MA, Feitosa VL. Venous drainage of the soleus muscle. Surgical and Radiologic Anatomy. 2008 Jun;**30**(4):341-345. PMID: 18330491. DOI: 10.1007/s00276-008-0333-6

[28] Kageyama N, Ro A, Tanifuji T, Fukunaga T. Significance of the soleal vein and its drainage veins in cases of massive pulmonary thromboembolism. Annals of Vascular Diseases. 2008;**1**(1):35-39. PMID: 23555336. DOI: 10.3400/avd.AVDoa07004. Erratum in: Ann Vasc Dis. 2008;**1**(2):122

[29] Ro A, Kageyama N. Clinical significance of the soleal vein and related drainage veins, in calf vein thrombosis in autopsy cases with massive pulmonary thromboembolism. Annals of Vascular Diseases. 2016;**9**(1):15-21. PMID: 27087868. DOI: 10.3400/avd.oa.15-00088

[30] Ohgi S, Ohgi N. Relationship between specific distributions of isolated soleal vein thrombosis and risk factors. Annals of Vascular Diseases. 2014;**7**(3):246-255. PMID: 25298825. DOI: 10.3400/avd.oa.14-00077

[31] Keijsers JM, Leguy CA, Huberts W, Narracott AJ, Rittweger J, van de Vosse FN. A 1D pulse wave propagation model of the hemodynamics of calf muscle pump function. International Journal for Numerical Methods in Biomedical Engineering. 2015 Jul;**31**(7):e02716. PMID: 25766693. DOI: 10.1002/cnm.2714

[32] Williams KJ, Ayekoloye O, Moore HM, Davies AH. The calf muscle pump revisited. Journal of Vascular Surgery: Venous and Lymphatic Disorders. 2014 Jul;**2**(3):329-34. DOI: 10.1016/j.jvsv.2013.10.053

Cardiac Anatomy for the Electrophysiologist with Emphasis on the Left Atrium and Pulmonary Veins

Gabriel Cismaru, Lucian Muresan, Puiu Mihai,
Radu Rosu, Gabriel Gusetu, Andrei Cismaru,
Dana Pop and Dumitru Zdrenghea

Additional information is available at the end of the chapter

http://dx.doi.org/10.5772/intechopen.69120

Abstract

This chapter aims to provide basic anatomical knowledge for the interventional electrophysiologists to understand catheter placement and ablation targets. We begin with the location of the heart inside the mediastinum, position of cardiac chambers, pericardial space and neighboring structures of the heart. We continue with the right atrium and important structures inside it: sinus node, cavotricuspid isthmus, Koch's triangle and interatrial septum with fossa ovalis. A special part of this chapter is dedicated to the left atrium and pulmonary veins with the venoatrial junction, important structures for catheter ablation of atrial fibrillation. We finish our description with both ventricles with outflow tracts and the coronary venous system.

Keywords: catheter ablation, cavotricuspid isthmus, anatomy, dissection, heart chambers

1. Introduction

The recent development of catheter ablation was possible, thanks to a rigorous understanding of cardiac anatomy. Appropriate cardiac structure knowledge is relevant to avoid or minimize complications during catheter placement and RF application. New strategies for pulmonary vein isolation appeared and made the procedure safety and efficient, after a meticulous characterization of the atrial muscular sleeves that prolong inside the veins.

This chapter aims to provide basic anatomical knowledge for the interventional electrophysiologists to understand catheter placement and ablation targets. We begin with the location of the heart inside the mediastinum, position of cardiac chambers, pericardial space and neighboring structures of the heart. We continue with the right atrium and important structures inside it: sinus node, cavotricuspid isthmus, Koch's triangle and interatrial septum with fossa ovalis. A special part of this chapter is dedicated to the left atrium and pulmonary veins with the venoatrial junction, important structures for catheter ablation of atrial fibrillation. We finish our description with both ventricles with outflow tracts and the coronary venous system.

2. General anatomy of the heart

The heart is positioned 2/3 to the left and 1/3 to the right of the midline of the thorax, between the two lungs. The anterior part of the heart consists of the right ventricle, which lies behind the sternum (**Figure 1**). The base of the heart lies in front of the spine. Neighboring structure is separated from the heart by the pericardium (**Figure 2**).

The posterior wall of the left atrium comes in contact with the esophagus [1], which can be close to the right or left orifices of the pulmonary veins. Catheter ablation at this level should be performed with lower energy or with temperature monitoring to avoid the risk of atrio-esophageal fistula.

On the outer surface of the pericardium [2] descend the right and left phrenic nerves. The right nerve is close to the superior vena cava and right superior pulmonary vein (RSPV) and can be damaged during cryoablation of the RSPV or RF ablation near the sinus node. In order to avoid the damage of the right phrenic nerve, high-output stimulation is performed at the level of superior vena cava and right atrium to distinguish hiccups and avoid ablation lesions at this level. The left phrenic nerve [3] is in the proximity of the left atrial appendage and can be damaged when ablating at the base of the appendage, especially in patients with persistent atrial fibrillation when extensive ablation is needed [4].

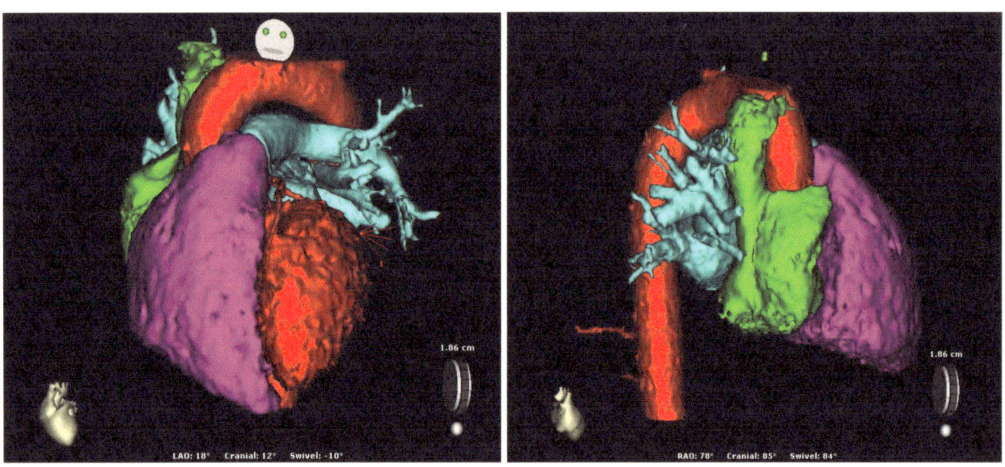

Figure 1. Heart chambers as seen in computed tomography.

http://dx.doi.org/10.5772/intechopen.69120

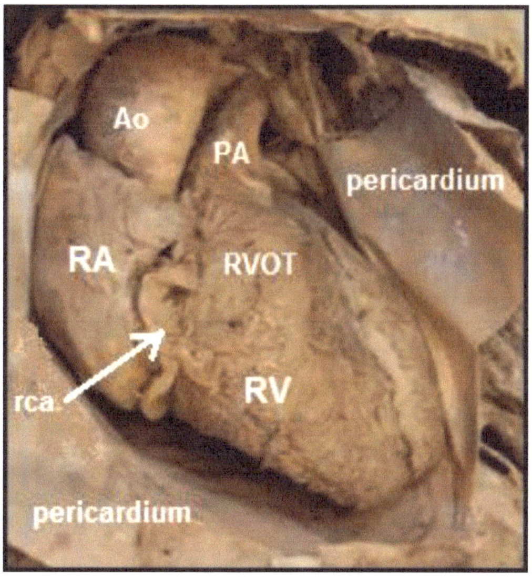

Figure 2. Parietal pericardium separating the heart from other mediastinal structures. RVOT = right ventricular outflow tract, Ao = Aorta, RA = right atrium, RV = right ventricle; rca = right coronary, PA = pulmonary artery.

3. The right atrium

The right atrium has four classical components: the vestibule, the venous part, the right atrial appendage and the interatrial septum (**Figure 3**). The sinus node is located [5] on the anterolateral part of the right atrium, at the level of the cavoatrial junction with superior vena cava. The right atrial appendage has prominent muscular bundles that give the high amplitude potential recorded at this level with a diagnostic catheter. The vestibule surrounds the orifice of the tricuspid valve and has a smooth appearance, without pectinate muscles, present in other regions of the right atrium. The venous component lies between the superior and the inferior vena cava and forms the posterior aspect of the right atrium. It has a smooth wall that is separated from the pectinated atrial zone by the terminal crest.

The terminal part of the crest divides into small muscular bundles that form the cavotricuspid isthmus [6], an important region between the inferior vena cava and the vestibule of the tricuspid valve that is "burned" during RF ablation of typical atrial flutter. The inferior vena cava has a fibrotic partial valve that is called the Eustachian ridge, a thin flap that is an important marker for catheter ablation of atrial flutter. An important percentage of patients present pouches and recesses at the level of the cavotricuspid isthmus. These structures can make the ablation of typical atrial flutter more difficult.

The fibrotic prolongation of the Eustachian valve toward the septum is called the tendon of Todaro. With the septal leaflet of the tricuspid valve and the coronary sinus, orifice forms the triangle of Koch, an anatomical structure that every electrophysiologists should know. At this level, ablation of intranodal reentry is performed and sometimes accessory pathways can also be ablated at this level.

The interatrial septum is used to access the left atrium through transseptal puncture (**Figure 4**). The true septum that can be crossed with a transseptal needle is the fossa ovalis; at this level, the puncture is safe without the risk of pericardial bleeding.

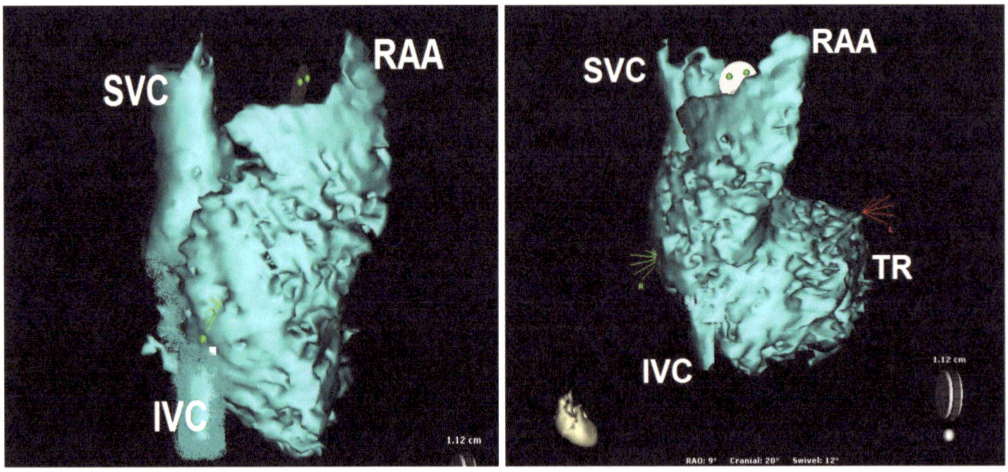

Figure 3. Computed tomography of the right atrium (the contrast substance from the cavity is subtracted and gives an image similar to an internal cast). (A) RAO view; (B) anterior view. The right atrial appendage (RAA) is an anterior structure. TR = tricuspid valve, IVC = inferior vena cava, SVC = superior vena cava.

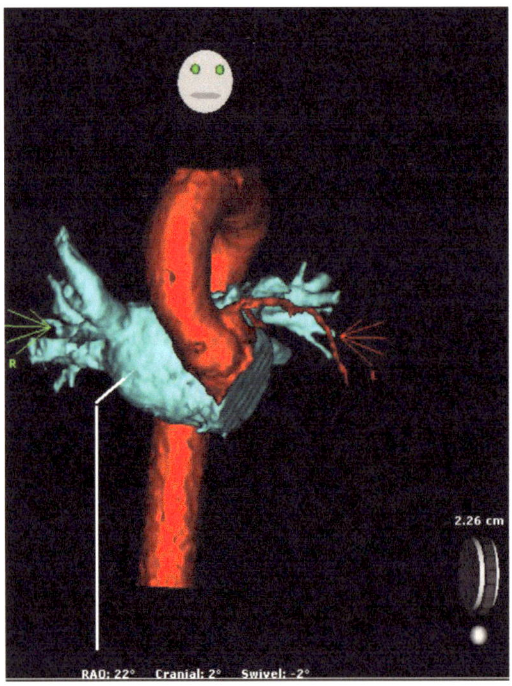

Figure 4. Position of the left atrium and ascending aorta; transseptal needle on the intratrial septum.

When a patent foramen ovale (PFO) is present, the electrophysiologists are tented to use it as a way to the left atrium. Owing to its location, PFO directs the catheter toward the anterior and superior wall of the left atrium, making difficult the ablation of the right pulmonary veins. When using the PFO, the electrophysiologist should know that the risk of roof perforation is higher.

Catheters positioned inside the right atrium facilitates the understanding of the activation sequence in different types of arrhythmias: A Halo catheter with 20 pols positioned along the tricuspid valve records the electrical signals of counterclockwise activation in case of a typical atrial flutter; a circumferential mapping catheter placed at the base of the superior vena cava records the electrical activity when mapping sinus node reentrant tachycardia.

4. The left atrium

The left atrium is the most posterior cardiac chamber. Behind the LA lies the tracheal bifurcation, the esophagus, the descending thoracic aorta and more posteriorly the vertebral column.

The left atrium is a structure composed of four parts (**Figure 5**): the venous component that receives the pulmonary veins, the left atrial appendage, the vestibule of the mitral valve and the left interatrial septum [7]. The walls of the left atrium are anterior, superior, left lateral, septal and posterior. The interatrial septum has a 45–60 degrees angulation to the horizontal plane. The superior and posterior walls of the left atrium are smooth, whereas the left appendage presents pectinate muscles. The left atrial appendage has a particular morphology described by Biase et al.: cactus-like 30%, chicken wing 48%, windsock 19% and cauliflower 3%. Patients with chicken wing morphology are less likely to develop thrombus at this level. The left appendage is smaller than the right one [8].

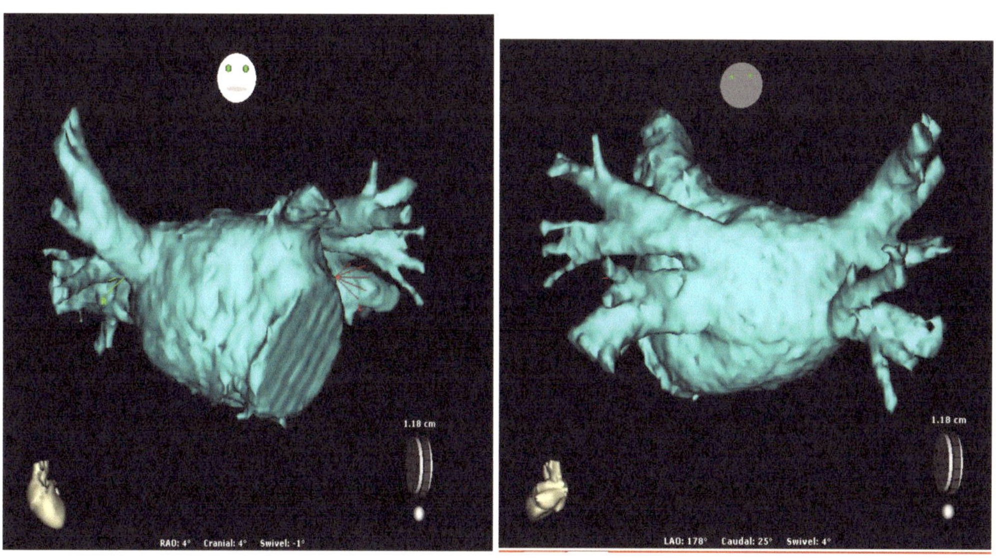

Figure 5. The left atrium with the four pulmonary veins: anterior and posterior views.

The transverse diameter of the left atrium is the largest because left atrium lies between the ascendant aorta anteriorly and the spine posteriorly and the dilation of the cavity is made between these structures. The roof of the left atrium is close to the right pulmonary artery and the bifurcation of the pulmonary trunk [9].

Between the two atrial chambers left and right, there are muscular bridges made of atrial myocardium. The most important is the Bachmann bundle, which is composed of parallel myocardial strands, extending from the left appendage to the right appendage.

Atrial fibrillation results in remodeling of the left atrium with dilation and fibrosis (the so-called atrial cardiomyopathy).

The left atrial isthmus, which is not a distinct anatomical structure, is the connecting line between the inferior margin of the LIPV and the mitral annulus. The line is used when ablating persistent atrial fibrillation and increases the success rate of the technique.

5. The pulmonary veins

The pulmonary veins drain oxygenated blood from the lungs to the posterior aspect of the left atrium. The left pulmonary veins ostia are located more superiorly than the right ostia. The right and left superior pulmonary veins are anterior and superior structures, whereas the right and inferior pulmonary veins are posterior and downwards [10]. The orifices of the veins are oval in shape with a superoinferior diameter longer than anteroposterior diameter. Usually, there are two veins on the right and two on the left side but sometimes supplementary veins can be found, more frequently on the right side. A frequent anatomical variation is the presence of a common trunk on the left side.

The musculature of the left atrium extends inside the pulmonary veins developing muscular sleeves; the longest being found in the superior veins: LSPV and RSPV. The myocardial fibers extend at a length of 1–3 cm. Usually, the sleeves are more important on the inferior part of the superior veins and on the superior part of the inferior veins.

The superior pulmonary vein is separated from the left atrial appendage by the left ridge, which is a structure that needs to be ablated during RF ablation of atrial fibrillation because muscular sleeves are very well developed at this level. The most challenging part is to obtain a good contact with the ridge.

There is a direct link between the pressure inside the pulmonary veins and abnormal electrical activations from the vein. When the left atrial pressure increases above 10 cm H2O, the junction between the LA and pulmonary veins becomes the source of abnormal activations from the pacemaker cells (Cajal-like cells).

The modern treatment of paroxysmal atrial fibrillation is pulmonary vein isolation (**Figure 6**) because ectopic triggers are found inside the pulmonary veins. Ablation of pulmonary foci is effective but with a high risk of pulmonary stenosis, therefore in the last years, ablation is performed at the level of venous antrum and aims to isolate the veins from the atrium.

http://dx.doi.org/10.5772/intechopen.69120

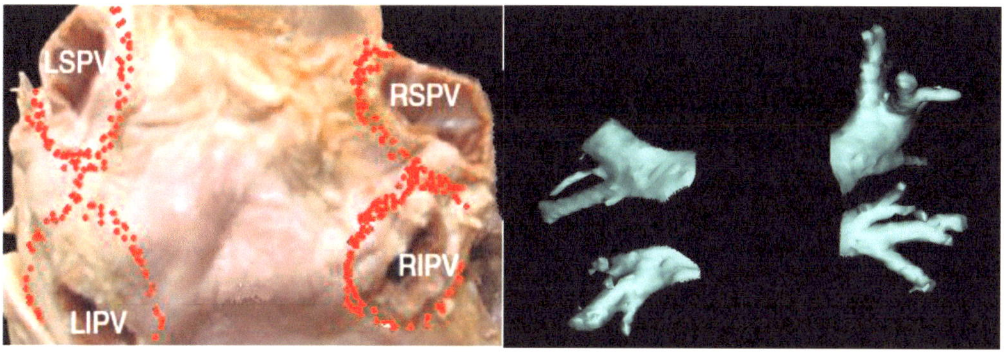

Figure 6. During catheter ablation for paroxysmal atrial fibrillation, pulmonary veins are isolated (it can be done either endocardial during electrophysiological study or epicardial during cardiac surgery).

6. The right ventricle

The right ventricle is the heart chamber that is situated the most anterior (**Figure 7**). It has three portions: the outlet or RVOT [11] that is continued with the pulmonary artery, the inlet which is delimitated by the tricuspid valve and papillary muscles, and the apical part (**Figure 8**). A thick moderator band can be present inside the right ventricle, making catheter manipulation difficult inside the RV [12].

The RVOT is superior to the left ventricular outflow tract which crosses the RVOT in a posterior position. The myocardium of RVOT is very thin, and perforations can result when a stiff tip ablation catheter is advanced directly to RVOT.

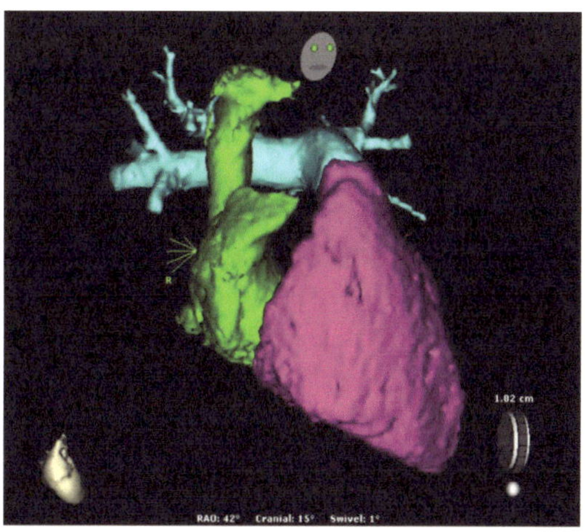

Figure 7. Right heart chambers: right ventricle, right atrium and pulmonary artery.

The RVOT is the most frequent region of benign monomorphic premature ventricular complexes. Ablation is carried at this level using pacemapping and activation mapping at the level of septal, lateral, anterior and posterior RVOT. This structure is in close relation with the left ventricular outflow tract and also aortic cusps, and sometimes mapping of these structures should be performed when ablation is not effective in the RVOT.

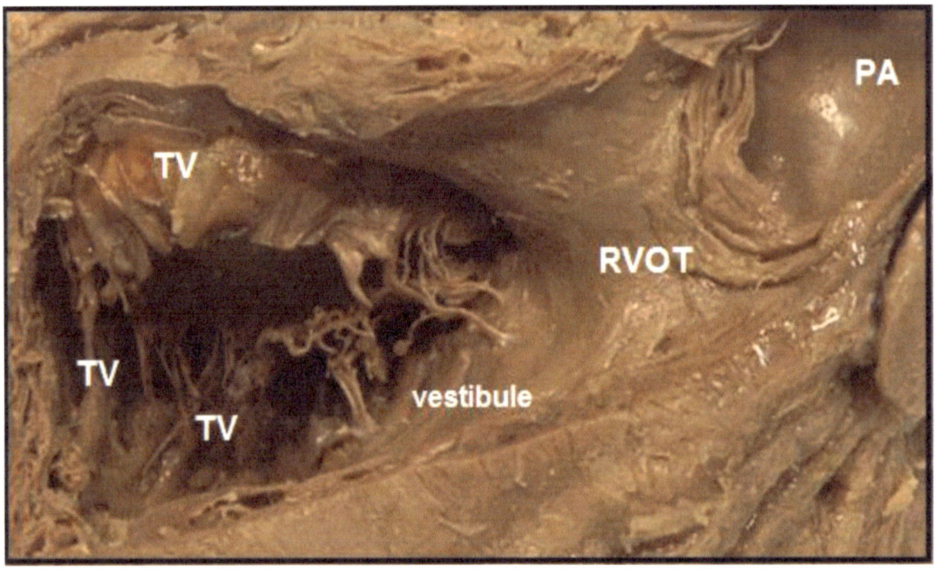

Figure 8. Inlet and outlet aspects of the right ventricle. Please note the smooth space near the tricuspid valve, which is called tricuspid valve vestibule. PA = pulmonary artery, RVOT = right ventricular outflow tract, TV = tricuspid valve.

7. The left ventricle

The left atrium is continued by the mitral valve and the left ventricle (**Figure 9**) which also has three components: inlet, outlet and apical part. The apical LV extends from the insertion of the papillary muscles to the apex. The walls of the left ventricle are thicker than those of the right ventricle but the trabeculations are finer than those of RV [13]. The anterosuperior papillary muscle and the posteroinferior papillary muscle (**Figure 10**) can be sources of ventricular premature contractions that have to be differentiated from PVC arising from the left bundle conduction system.

The LVOT is directed superiorly and anteriorly. It can be a source of ventricular premature contractions. When mapping the left ventricle, access can be achieved through the transmitral anterograde approach that requires a transseptal puncture or through the transaortic retrograde approach.

http://dx.doi.org/10.5772/intechopen.69120

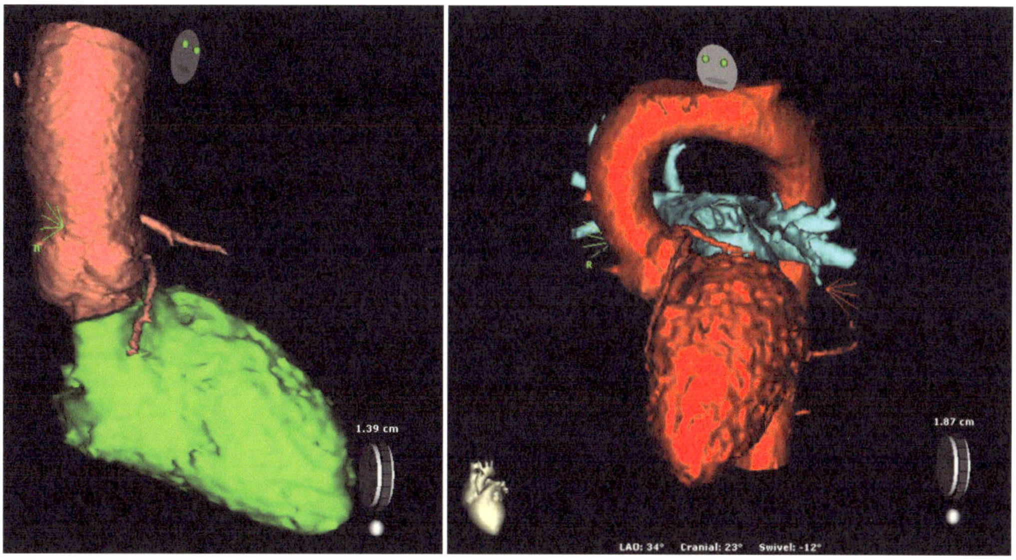

Figure 9. (A) The left ventricle with the ascendant aorta from a RAO view. (B) The left ventricle with the ascendant aorta and left atrium as seen from LAO.

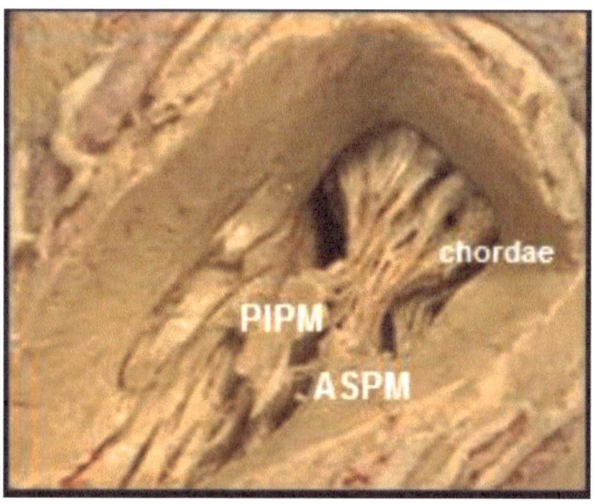

Figure 10. Anterosuperior (ASPM) and posteroinferior papillary muscles (PIPM) with chordae to the mitral valve.

8. The coronary veins

Most of the venous flow of the heart is collected by the coronary venous system. The coronary sinus drains the great cardiac vein and the middle cardiac vein as well as other small veins (**Figure 11**). The cardiac veins might be used for catheter ablation of ventricular premature contractions or ventricular tachycardias. Electrophysiologists can reach the epicardium of the left ventricle through the venous system (**Figure 12**). Small diagnostic and therapeutic (2F–5F) catheters are used inside the coronary veins, and ablation is performed usually with irrigated catheters to avoid perforation of outer walls that are not protected by muscular bundles.

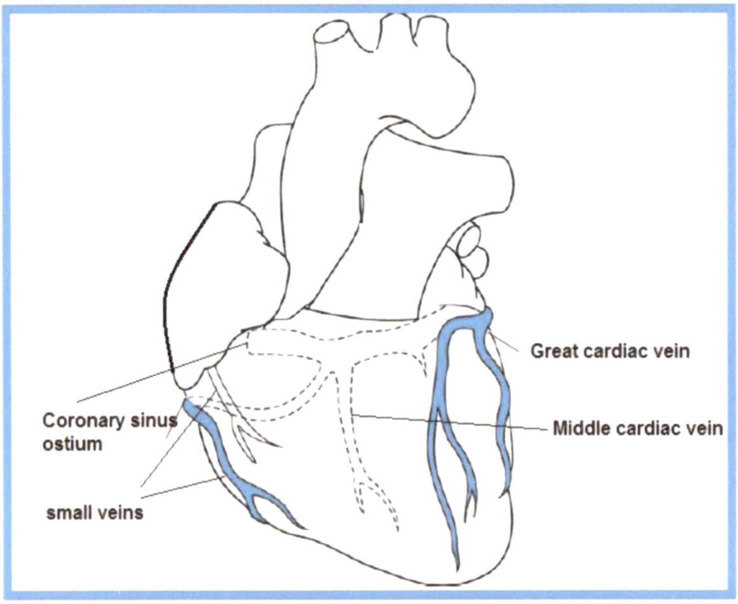

Figure 11. Coronary venous system of the heart.

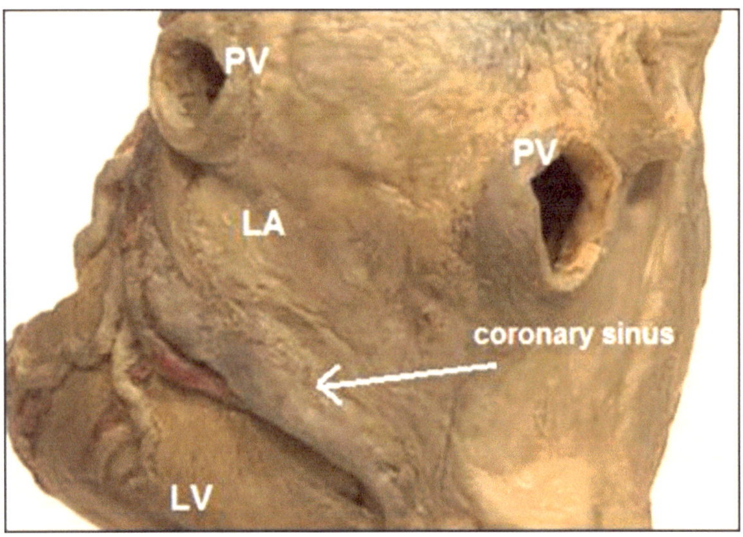

Figure 12. The coronary sinus drains venous blood from the cardiac veins. It is located between the left atrium and left ventricle. LA = left atrium, LV = left ventricle; PV = pulmonary vein.

Sometimes muscular bundles form sleeves that cover the coronary sinus and also prolong into the left atrium. These muscular bundles are target of ablation in patients with persistent atrial fibrillation that need substrate modification.

Sometimes the middle cardiac vein is dilated and forms a diverticulum of the coronary sinus. Catheter ablation might be performed at this level as posteroseptal accessory pathways could be located at this level.

The coronary sinus orifice is bordered by a small flap: the Thebesian valve that is easily passed by a diagnostic catheter because this valve is incomplete.

9. Conclusions

Our understanding of cardiac anatomy has grown exponentially in the era of catheter ablation. The knowledge of the anatomy of a specific cardiac chamber and its relationship with neighboring structures is relevant for interventional electrophysiologists when mapping and ablating different arrhythmias.

Each cardiac structure can be a source of arrhythmias, and the knowledge of the particular anatomy facilitates the understanding of the mechanism behind the abnormal rhythm and how it can be controlled. Safe ablation comes not only from an improved understanding of the gross cardiac anatomy but also from a good awareness of the histological characteristics and architectural microstructure.

Acknowledgements

The research was supported by the internal grant of the University of Medicine and Pharmacy Iuliu Hatieganu Cluj-Napoca No 4994/1/08.03.2016.

Author details

Gabriel Cismaru[1]*, Lucian Muresan[1], Puiu Mihai[1], Radu Rosu[1], Gabriel Gusetu[1], Andrei Cismaru[2], Dana Pop[1] and Dumitru Zdrenghea[1]

*Address all correspondence to: gabi_cismaru@yahoo.com

1 5th Internal Medicine Department, Cardiology-Rehabilitation, Iuliu Hatieganu University of Medicine and Pharmacy, Cluj-Napoca, Romania

2 11th Oncology Department, Oncology, Iuliu Hatieganu University of Medicine and Pharmacy, Cluj-Napoca, Romania

References

[1] Sanchez-Quintana D, Cabrera JA, Climent V, et al. Anatomic relations between the esophagus and left atrium and relevance for ablation of atrial fibrillation. Circulation. 2005;**112**:1400-1405

[2] D'Avila A, Scanavaca M, Sosa E, et al. Pericardial anatomy for the interventional electrophysiologists. Journal of Cardiovascular Electrophysiology. 2003;**21**:339-379

[3] Sanchez-Quintana D, Cabrera JA, Climent V, et al. How close are the phrenic nerves to cardiac structures ? Implications for cardiac interventionists. Journal of Cardiovascular Electrophysiology. 2005;**16**:309-313

[4] Buch E, Vaseghi M, Cesario DA, et al. A novel method to prevent phrenic nerve injury during catheter ablation. Heart Rhythm. 2007;**4**:95-98

[5] Sanchez-Quintana D, Anderson RH, Cabrera JA, et al. Sinus node revisited in the era of electroanatomical mapping and catheter ablation. Heart. 2005;**91**:189-194

[6] Cabrera JA, Sanchez-Quintana D, Farre J, et al. Angiographic anatomy of the inferior right atrial isthmus in patients with and without history of common atrial flutter. Circulation. 1999;**99**:3017-3023

[7] Siew YH, Cabrera HA, Sanchez-Quintana D. Left atrial anatomy revisited. Circulation: Arrhythmia and Electrophysiology. 2012;**5**:220-228

[8] Sanchez-Quintana D, Lopez-Minguez JR, Macias Y, Cabrera JA, Saremi F. Left atrial anatomy relevant to catheter ablation. Cardiology Research and Practice. 2014;1-17:289720

[9] Ho SY, McCarthy KP, Faletra FF. Anatomy of the left atrium for interventional echocardiography. European Journal of Echocardiography. 2011;**12**:i11-i15

[10] Klimek-Piotrowska W, et al. Normal distal pulmonary vein anatomy. PeerJ4. 2016;**e1597**: 1-13.

[11] Asirvatham SJ. Correlative anatomy for the invasive electrophysiologists: Outflow tract and supravalvular arrhythmia. Journal of Cardiovascular Electrophysiology. 2009;**20**:955-968

[12] Ho SY, Nihoyannopoulos P. Anatomy, echocardiography and normal right ventricular dimensions. Heart. 2006;**92**:i2-i13

[13] Abouezzeddine O, Suleiman M, Buescher T, Kapa S, Friedmn PA, Jahangir A, et al. Relevance of endocavitary structures in ablation procedures for ventricular tachycardia. Journal of Cardiovascular Electrophysiology. 2010;**21**:254-254

Chapter 7

Anatomical, Biological, and Surgical Features of Basal Ganglia

Nuket Gocmen Mas, Harun Muayad Said,
Murat Tosun, Nilufer Yonguc, Yasemin Soysal and
Hamit Selim Karabekir

Additional information is available at the end of the chapter

http://dx.doi.org/10.5772/intechopen.68851

Abstract

Basal ganglia refers to the deep gray matter masses on the deeply telencephalon and encompasses a group of nuclei and it influence the information in the extrapyramidal system. In human they are related with numerous significant functions controlled by the nervous system. Gross anatomically, it is comprised of different parts as the dorsal striatum that are consisted of the caudate nucleus and putamen and ventral striatum which includes the nucleus accumbens, olfactory tubercle, globus pallidus, substantia nigra, and subthalamic nucleus. Nucleus accumbens, is also associated with reward circuits and has two parts; the nucleus accumbens core and the nucleus accumbens shell. Neurological diseases are characterized through the obvious pathology of the basal ganglia, and there are important findings explaining striatal neurodegeneration on human brain. Some of these diseases are induced by bacterial and/or viral infections. Surgical interference can be one alternative for neuronal disease treatment like Parkinson's Disease or Thiamine Responsive Basal Ganglia Disease or Wilson's Disease, respectively in addition to the vascular or tumor surgery within this area. Extensive knowledge on the morphological basis of diseases of the basal ganglia along with motor, behavioral and cognitive symptoms can contribute significantly to the optimization of the diagnosis and later patient's treatment.

Keywords: anatomy, biology, surgery, basal ganglia

1. Structure and function of human basal ganglia

The term "basal ganglia" refers to the deep gray matter masses on the deep telencephalon and encompasses a group of nuclei [1]. Generally, basal ganglia influence information in the extrapyramidal system and in human beings they are related with numerous significant functions controlled by the nervous system, including control of the voluntary motor movements, procedural learning, and habitual behaviors such as eye movements, cognition, and emotions. The nuclei group is placed deep beneath the cortical area of the brain.

Additionally, the basal ganglia specialize in processing data on movement and in fine adjustment of the brain circuit activity that defines the best suitable response in specific habitual conditions/actions such as riding a bicycle, playing a piano, and so on. They also play a major role while planning movement and learning novel actions in new situations [2, 3]. From an embryological point of view, the central nervous system (CNS) develops in early stages of embryological development in many of mammals, including humans, from the neural plate. In humans, at the middle of the third week of development, ectodermal cells from epiblast fold to form a neural groove and then differentiate into the neural tube. At the 24th day, the anterior neuropore and then at the 27th day the posterior neuropore are closed [4–6], while the cerebrospinal fluid is secreted by the ependymal cells, which are lining within the ventricular system, prosencephalic, mesencephalic, and rhombencephalic vesicles that develop from the rostral part of the neural tube and also the spinal cord from the caudal part of the neural tube. While telencephalic vesicles developed from a part of the prosencephalic vesicle at the late 34th day, at the 36th day, anlage primordium of the future cerebral cortex, basal ganglia, and olfactory bulb can be clearly identified. At the eighth week of the development of the basal ganglia, neuroepithelium is clearly defined near the other epithelial structures such as thalamic, hippocampal, and hypothalamic epitheliums. Basal ganglia neurons are derived from this epithelium [4–6]. Between the 19 and 22 gestational weeks, the neuroepithelial cell layer becomes thicker and new neurons are generated, which migrated into the striatum. Between the 23 and 28 gestational weeks, in basal ganglia, the neuroepithelial cell layer over some places of caudate nucleus became faint. Again, at week 27 of the development in the basal ganglia, neuroepithelial cellularity is scant. Between the 29 and 33 gestational weeks in the basal ganglia, neuroepithelium was thicker over the striatum and nucleus accumbens. At the 32nd week of the development process, the number of the glial cells increased [4–6]. On the other hand, during this period, the onset of myelinization was very silent. At the 35th week, it can be detected microscopically in the subventricular zone, neuroepithelial cells are limited and the glial cell number is increased. In addition, there is a slight onset of myelinization in the internal capsule. At term, proliferating neuroepithelial cells can be detected in the subventricular zone, while the numbers of astrocytes decrease in the internal capsule and basal ganglia. In the postnatal period, basal ganglia and diencephalic neurons were well organized, and myelinization of the internal capsule is complete approximately at the 6th week postnatally. Two years postnatal, in basal ganglia, all of the mature histological structures can be clearly detected in the caudate and putamen, and myelinization of internal capsule appeared completely [4–6].

In our investigation, we aimed to highlight the anatomical, biological, and surgical importance of cortical-basal ganglia circuits and their role in the pathogenesis of neurological process. Depending on the facts available nowadays and our experience, we developed an opinion that detailed anatomy related information in embryology, histology, and gross anatomy as well as molecular and surgical information on the basal ganglia and neighborhood structures may cause confused clinical outcomes and possibly the option of renovating the morphological brain structure after intervention to the region of intervention [4–6].

http://dx.doi.org/10.5772/intechopen.68851

Gross anatomically, the basal ganglia are composed of different parts including the dorsal striatum consisting the caudate nucleus and putamen and ventral striatum, which includes the nucleus accumbens, olfactory tubercle, globus pallidus, substantia nigra, and subthalamic nucleus. Each of these parts possesses defined, complex internal morphological and biological features. Basal ganglia are composed of several subcortical nucleus groups that are located deep on each of the cerebral hemispheres. They are called "lentiform nucleus," which includes both the putamen and globus pallidus; the "striatum," which includes the nucleus caudatus and putamen; and "corpus striatum," including the caudate nucleus and the nucleus lentiformis, and the others including the claustrum, subthalamic nucleus, nucleus accumbens, and their projections. The substania nigra is also a basal nucleus, which is placed on the mesencephalon [4–6]. This nuclear group carries heterogeneous formations, functionally, terminologically and phylogenetically. Details are presented in **Figure 1**.

According to the previous classification, the amygdaloid body was considered to be a part of the basal ganglia (archistriatum). Due to the acquisition of new scientific data associated with it, anatomists considered it as a functional educative part [8].

Clastrum is located on the lateral to the putamen and medial to the insula, which was noted by some sources to be a part of the basal ganglia [8]. The globus pallidus externus was originally revealed as a simple relay within the basal ganglia [9].

The nucleus accumbens is also associated with reward circuits that are located in the basal forebrain region superior to the preoptic region of the hypothalamus, while the prefrontal area was on both cerebral hemispheres, whose mission is planning and motivating movement performed by the body. The nucleus accumbens has two parts: the nucleus accumbens core and the nucleus accumbens shell, which consist of their own morphology and functions [10–15].

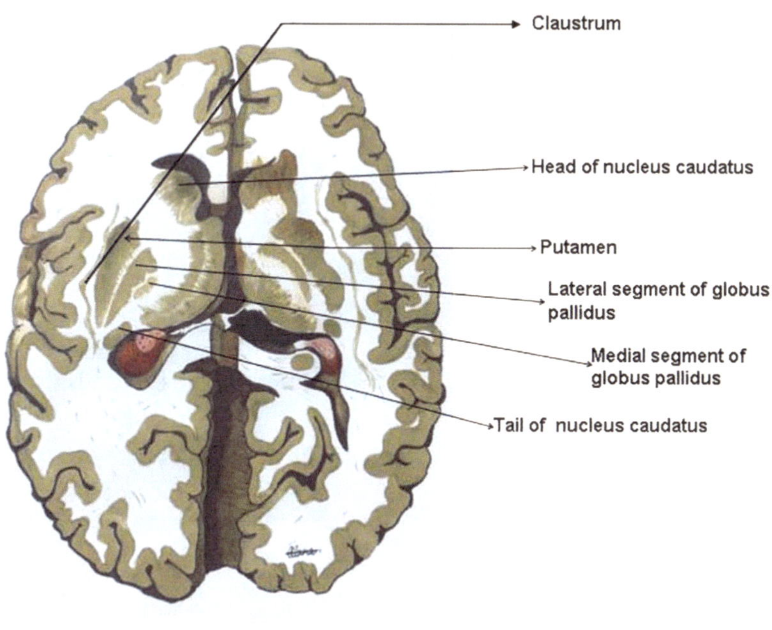

Figure 1. Nuclei of the basal ganglia showing the different components [7], with modifications.

From the histophysiological point of view, basal ganglia circuits contain an assortment of cell types that mediate synaptic interactions within and between basal ganglia nuclei [16]. Neurotransmitters also play an important role in the different areas of the basal ganglia. For example, dopamine, which has a very important function within the basal ganglia, is the source of the striatal input in the substantia nigra (**Figure 2**).

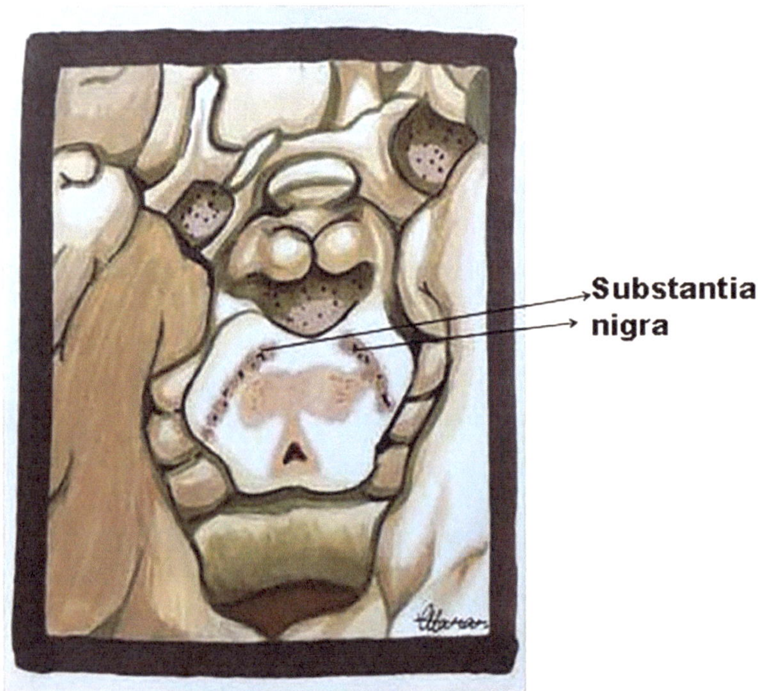

Figure 2. Illustrations showing the substancia nigra, according to [8] with modifications.

2. Pathogenic conditions of the basal ganglia

Huntington' and Parkinson's diseases are caused by the degeneration of dopamine-producing cells in the substantia nigra [4, 17]. On the other hand, most of the neurons in the basal ganglia use *gamma*-Aminobutyric acid (GABA) as a neurotransmitter, which possesses an inhibitory effect on the target neurons. Acetylcholine is another important neurotransmitter and is regularly used by both external inputs to the striatum and by a group of striatal neurons. Although the total number of cholinergic neurons is the smallest in all brain neurons, one of the major acetylcholine concentration regions is the striatum [4, 17].

Striatum (**Figures 3** and **4**) is currently considered to be the largest region of the basal ganglia that arise from numerous large and small bundles of nerve fibers [18]. The histological organization of the striatum is considered very complex. The great populations of striatal neurons

http://dx.doi.org/10.5772/intechopen.68851

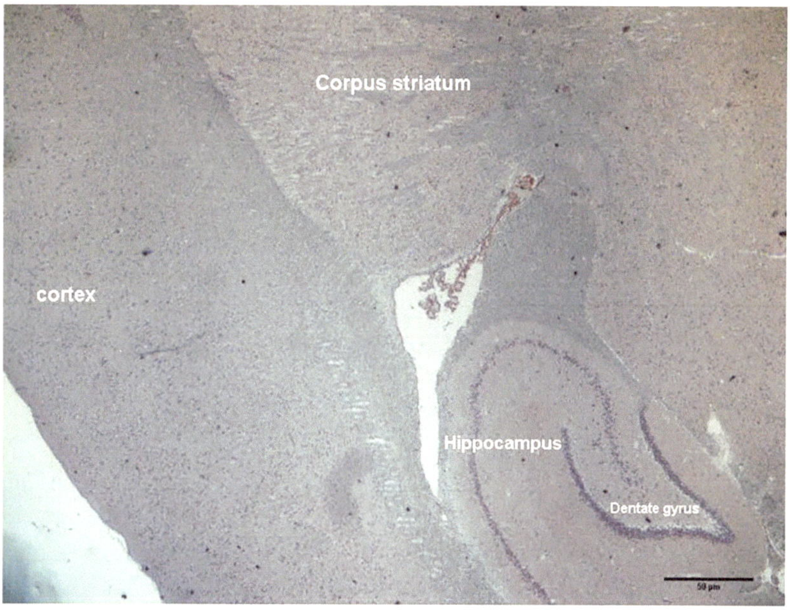

Figure 3. Corpus striatum in the sagittal section stained with caspase 3 immunostaining. Magnification, ×2 (with courtesy to Esra ASLAN MD).

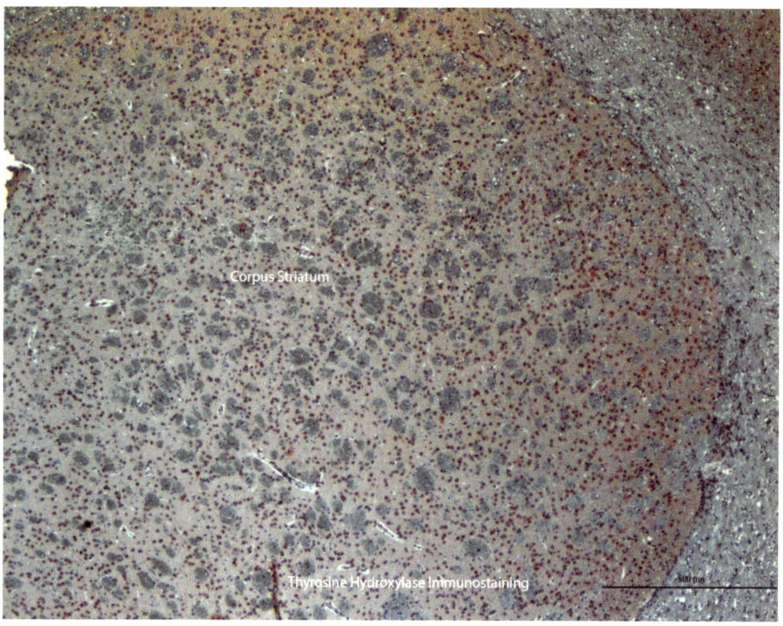

Figure 4. Corpus striatum in the coronal section with thyrosine hydroxylase immunostaining. Magnification, ×10. (with courtesy to Esra ASLAN, MD, Afyon Kocatepe University).

are medium spiny neurons, which are GABAergic cells with small bodies, densely covered dendritic spines that receive input from the cortex and the thalamus. The cholinergic neuron population in striatum comprises of the cholinergic neurons with smooth dendrites [2].

The pallidum consists of both the globus pallidus and the ventral pallidum. The globus pallidus can be divided functionally into the internal and external segments. Histologically, both segments have primarily GABAergic neurons [19]. While external segments receive inputs mainly from the striatum and pass through the subthalamic nucleus, the internal segment receives input from direct and indirect pathways. Pallidal neurons functions basically via de-inhibition, a mechanism in which there are inhibitory effects on the target [2].

The substantia nigra is a mesencephalic gray matter portion of the basal ganglia, which is divided into two parts: pars compacta and pars reticulate. Although pars compacta produces dopamine, which plays a major role as a regulator neurotransmitter in the striatal pathway, the pars reticulate has inhibitory effects on the thalamus [19].

The subthalamic nucleus is a diencephalic gray matter portion of the basal ganglia and produces glutamate, which is an excitatory neurotransmitter in the ganglia. This nucleus, while receiving inhibitory input from external part of the globus pallidus, also sends excitatory input to the internal part of the globus pallidus.

3. Basal ganglia-related pathological conditions

Clinically, many neurological diseases are characterized through the obvious pathology of the basal ganglia, and there are important findings that explain striatal neurodegeneration on the human brain. Some of these diseases are induced by bacterial and/or viral infections where the bacteria and/or virus introduces some genetic material and, as a consequence, either activated or downregulated some of the life's essential processes [20–26] or other diseases that are affected by the cytokine regulation in association with neurodegenerative diseases like TGF-β [27] or TNF-α [28] in addition to applications related to medical solutions like in dentistry [29]. Here, and as a consequence, affecting either the functionality or the neuronal structure or both of them can be seriously affected. Lately, more knowledge about the problems of the basal ganglia patients with neurodegenerative, vascular, metabolic, inflammatory, immunologic, allergic, congenital, traumatic, endocrine, malignant, and neurophyschiatric diseases became available [30–32]. A comprehensive understanding of the striatal projection loss while receiving striatal input/output on the neurons will contribute to the available knowledge related to the pathogenesis of the neurological diseases. Surgical interference can also be one alternative for neuronal disease treatment as it is the case for Parkinson's disease, thiamine responsive basal ganglia disease or Wilson's disease, respectively, in addition to the vascular or tumor surgery within this area. The lesions of the basal ganglia can cause tremors, grimaces, and repetitive movements [4, 17]. At the same time, in different pathological processes, such as Kernicterus, Tourette syndrome, hemiballismus, obsessive-compulsive disorder, neonatal and lacunar infarction, Huntington's and Parkinson's diseases, basal ganglia neurons were affected. Again, in carbon monoxide poisoning, selective necrosis is caused in the globus palidus [3, 33].

http://dx.doi.org/10.5772/intechopen.68851

4. Potential approaches for basal ganglia disease treatment

In order to limit or inhibit this type of disorders, gene therapeutic [34] based treatment modalities bear potential for the treatment of nervous system diseases or disorders, these include viral vector systems [35, 36], gene-based vaccines and immunotherapy [37, 38], plasmid DNA applications [39], cytokine targeting like TNF-α targeting [28], epigenetic targeting [40] and anti nervous system degenerative diseases treatment by molecular regulators RNAi applications [41]. Several studies on basal ganglia supported by data aligned to age- and/or gender-dependent relation of intelligence with volumes of the nuclei were presented recently [42, 43], still limited results were known regarding the potential influence of age- and sexual distinctive diseases on the subcortical nuclei [44]. The basal ganglia (BG), which play a major role in selecting and shaping motor and cognitive behaviors, are significant for connection among forebrain nuclei [9].

Surgically, in some neurologic diseases, using deep brain stimulation (DBS), which is an implanted electrical device modulate for distinct targets at the brain, resulted in the symptomatic improvement of movement disorder, especially [45, 46] in both hyper- and hypokinetic movement disorders of the basal ganglia deep brain stimulation (DBS) that is considered highly effective. The clinical benefit of DBS is based on the experience with prior surgical ablative therapies for the disorders of these regions, and, in part, used by neurosurgeons decades ago. The most commonly DBS-treated conditions were and are Parkinson's disease and dystonia, which are treated by electrical or radiofrequency lesioning of that region before DBS [46–53]. The duration and temperature are important for both procedures. Applying electrical current the functions that were partially or totally lost due to nervous system disease or injury can be restored [54].

In stereotaxic surgery, some entry points described by the authors for nucleus accumbens [55] were as follows: 7–9 mm below the anterior commissure-posterior commissure (AC-PC) line, 19–23 mm prior to the midpoint, and 4–10 mm lateral to the median line. The original target is the core of the nucleus accumbens. For deep brain stimulation applications as refractory major depression, Tourette syndrome and obsessive-compulsive disorders the stereotactic coordinates were as follows: 4–4.5 mm ventral to the AC-PC plane, 1.5–2.5 mm anterior to the anterior border of the AC, and 6.5–8 mm lateral to the midline [56, 57].

In treating obsessive-compulsive disorders with accompanying major depression, Aouizerate et al. reported their experience in DBS targeting as the tips of the electrodes were situated 3.0 mm below the AC-PC line, 8.9 mm lateral to the AC-PC line, and 36.5 mm anterior to the PC on the right side, and 1.7 mm below and 7.6 mm lateral to the AC-PC line and 31.4 mm anterior to the PC on the left side [58].

The subthalamic nucleus (STN) of advanced Parkinson's disease patients undergoing deep brain stimulation application is a prominent target for treatment. In many patients, to identify significant target, microelectrode recording (MER) is used. Moran in a previous work showed that trajectories served as a training set and found the error in predicting the STN entry to be (mean ± SD) 0.18 ± 0.84, and 0.50 ± 0.59 mm for the STN exit point, which yields a 0.30 ± 0.28 mm deviation from the expert's target center by using MER [59]. In the correlation analysis, there was a negative correlation between right substancia nigra (SN) volume and unified Parkinson disease rating scale (UPDRS) score ($r = -0.466$, $p = 0.038$) and there was a tendency but not a significant correlation between the left SN volume and UPDRS score ($r = -0.443$, $p = 0.050$).

In a previous approach, it was shown that subjects suffering from Parkinson's disease showed a significant asymmetry between both the left and right SN, nucleus caudatus, and nucleus lentiformis volumes ($p = 0.001$, $p < 0.001$, $p = 0.044$), with taking into account that the control subjects also showed a significant asymmetry between the volumes of left and right SN, nucleus caudatus, and nucleus lentiformis ($p < 0.001$, $p = 0.003$, $p < 0.001$, respectively). Mean volume values for SN, nucleus caudatus, and nucleus lentiformis are shown in **Table 1** [60].

Further, in the same approach, the group was examined after subgrouping according to gender into male and female subgroups, as seen in **Table 2**.

In addition, during our experience using stereological methods on the basal ganglia volumetry on the right-handed patients with the Parkinson's disease, we found that the left basal ganglia structure was smaller than left ones. However, when we compared them with the control cases, only substantia nigra possessed a smaller volume. Also, evaluation of the basal ganglia and substantia nigra volume in Parkinson's disease (PD) patients revealed a significant atrophy in SN in comparison to the healthy age-matched control subjects. However, significant atrophy in nucleus lentiformis and nucleus caudatus was not found during the study. Basal ganglia and the SN are the regions with predominantly pathological changes in PD [61]. There are studies in the literature that examine the volumetric differences in basal ganglia and SN anatomy in PD; however, there is no study to our knowledge in the literature that evaluates the asymmetrical volume changes by using the stereological technique [62, 63]. Cavalieri's principle of stereological approaches through point counting is accomplished by overlying each selected section using a regular grid of test points that is randomly positioned [64]. The Cavalieri theorem of systematic sampling combined with point counting proved to be a reliable, simple, inexpensive, and efficient method for volume estimation in MRI [65], and this stereological approach can provide valuable information during the morphological changes evaluated during Parkinson's disease development.

Surgical procedures applied for this purpose can be variable. Gallina et al. [66] provided details of the surgical procedure for both caudate and putaminal tracks through a single frontal entry point for six patients, and for the following 10 procedures and they used two completely distinct routes, with two separate entry points, each for the nucleus caudatus and putamen, respectively.

	Parkinson patients	**Controls**	***P* value**
	Mean ± SD	**Mean ± SD**	
SN left	0.67 ± 0.16	0.78 ± 0.13	0.026
SN right	0.75 ± 0.17	0.92 ± 0.18	0.005
NC left	4.49 ± 0.50	4.42 ± 0.24	0.602
NC right	4.47 ± 0.54	4.60 ± 0.21	0.632
NL left	5.29 ± 0.57	5.25 ± 0.53	0.832
NL right	5.40 ± 0.54	5.52 ± 0.52	0.473

Table 1. Standard volumes of the substantia nigra, nucleus caudatus, and nucleus lentiformis and a group of Parkinson's disease patients compared to subjects of a healthy control group [59].

	Parkinson disease			Control		
	Mean ± SD			Mean ± SD		
	Left side	Right side	*P value*	Left side	Right side	*P value*
Male						
Substancia nigra	*n* = 13	*n* = 13	0.005	*n* = 10	*n* = 10	0.001
	0.69 ± 0.14	0.75 ± 0.16		0.77 ± 0.15	0.91 ± 0.16	
Nucleus caudatus	*n* = 13	*n* = 13	0.007	*n* = 10	*n* = 10	0.001
	4.56 ± 0.51	4.72 ± 0.60		4.40 ± 0.23	4.58 ± 0.25	
Nucleus lentiformis	*n* = 13	*n* = 13	0.515	*n* = 10	*n* = 10	0.001
	5.29±0.65	5.37 ± 0.57		5.24 ± 0.54	5.56 ± 0.44	
Female						
Substancia nigra	*n* = 7	*n* = 7	0.020	*n* = 10	*n* = 10	0.008
	0.63 ± 0.20	0.75 ± 0.2		0.77 ± 0.15	0.91 ± 0.16	
Nucleus caudatus	*n* = 7	*n* = 7	0.014	*n* = 10	*n* = 10	0.039
	4.36 ± 0.5	4.57 ± 0.45		4.40 ± 0.23	4.58 ± 0.25	
Nucleus lentiformis	*n* = 7	*n* = 7	0.092	*n* = 10	*n* = 10	0.001
	5.28 ± 0.43	5.45 ± 0.52		5.24 ± 0.54	5.56 ± 0.44	

Table 2. Substancia nigra, nucleus caudatus, and nucleus lentiformis volumes in male and female groups of Parkinson disease patients [60].

In surgical processes, surgeons use a stereotactic frame that helps to ensure optimal positioning of desired targets, or frameless stereotactic systems [67] or, alternatively, neuronavigation or electrophysiological mapping of the brain for lesioning-related basal ganglia and for obtaining the main target in a three-dimensional manner in addition to protect the surrounding neural tissue [53, 55, 57, 68, 69].

5. Conclusion

Extensive knowledge on the morphological basis of diseases of the basal ganglia along with motor, behavioral, and cognitive symptoms may significantly contribute to the optimization of both the diagnosis (especially anatomical and histological) and later treatment of the patients, especially patients suffering from neurodegenerative, vascular, metabolic, inflammatory, immunologic, allergic, congenital, traumatic, endocrine, malignant, and neurophyschiatric diseases, in order to at least delay the breakout or the pathogenic degenerative process related to the their disease and/or improve the life quality of the patients. Experimental set-ups dealing with this level of problems can provide us the necessary information for the treatment modalities applied in human therapy. Gene therapeutic approaches can be a future effective alternative for these classes of disease treatment.

Acknowledgements

The authors would like to thank the Dokuz Eyül University and the Afyon Kocatepe University for their kind support. Özge Yilmaz Kusbaci, M.D. and Esra ASLAN M.D. Histologist and Embryologist at Afyon Kocatepe University Faculty of the Medicine Department of Histology and Embryology for the provision of both the Corpus striatum in sagittal and the Corpus striatum in coronal sections. Also, we would like to thank Ms. Alara Karabekir for the graphic design. Finally, we would like to thank Mr. Ali Ege Mas from the Near Eastern University, Cyprus, for technical support during the different stages of the research.

Author details

Nuket Gocmen Mas[1], Harun Muayad Said[2], Murat Tosun[3], Nilufer Yonguc[1], Yasemin Soysal[2] and Hamit Selim Karabekir[4*]

*Address all correspondence to: hskarabekir@hotmail.com

1 Department of Anatomy, Faculty of Medicine, Graduate Institute of Health sciences, Dokuz Eylül University, Izmir, Turkey

2 Department of Molecular Medicine, Graduate Institute of Health Science, Dokuz Eylul University, Izmir, Turkey

3 Department of Histology and Embryology, Faculty of Medicine, Afyon Kocatepe University, Afyonkarahisar, Turkey

4 Department of Neurosurgery, Dokuz Eylül University, Izmir, Turkey

References

[1] Mills SE. Central nervous system. In: Histology for Pathologist. 3rd ed. Baltimore, MD: Lippincott William & Wilkins; 2007

[2] Stocco A, Lebiere C, Anderson JR. Conditional routing of information to the cortex: A model of the basal ganglia's role in cognitive coordination. Psychological Review. 2010;**117**(2):541-574

[3] Weyhenmeyer JA, Gallman EA. Rapid Review of Neuroscience. Philadelphia, PA: Mosby Elsevier; 2007. p. 102

[4] Carlson B. Nervous system. In: Human Embryology and Developmental Biology. 4th ed. Philadelphia, PA: Mosby Elsevier; 2009. pp. 239-272

[5] Moore KL, Persaud, TVN, Torchia MG. Nervous system. In: The Developing Human. 9th ed. Elsevier Saunders; 2013. pp. 389-426

[6] Schoenwolf GC, Bleyl SB, Brauer PR, Francis–West PH. Development of the Central Nervous System. In: Larsen's Human Embryology. 4th ed. New York/Edinburgh: Churchill Livingstone; 2009. pp. 247-290

[7] Sandring S. Gray's Anatomy. 39th ed. New York/Edinburgh: Churchill Livingstone; 2005. pp. 419-430

[8] Netter FH. Atlas of Human Anatomy. 6th ed. Saunders; 2014. p. 111

[9] Saunders A, Huang KW, Sabatini BL. Globus pallidus externus neurons expressing parvalbumin interconnect the subthalamic nucleus and striatal interneurons. PLoS One. 2016 Feb 23;**11**(2):e0149798

[10] Albin RL, Young AB, Penney JB. The functional anatomy of basal ganglia disorders. Trends in Neurosciences. 1989;**12**:366-375

[11] DeLong MR. Primate models of movement disorders of basal ganglia origin. Trends in Neurosciences. 1990;**13**:281-285

[12] Carlson NR. Physiology of Behavior. 11th ed. Boston: Pearson; 2013. Print

[13] Wenzel JM, Rauscher NA, Cheer JF, Oleson EB. 2015 A role for phasic dopamine release with in the nucleus accumbens in encoding aversion: A review of the neurochemical literature. In: Carlson, NR, editor. Physiology of Behavior. 11th ed. Boston: Pearson; 2013. Print

[14] Ahsan RL, Allom R, Gousias IS, Habib H, Turkheimer FE, Free S, Lemieux L, Myers R, Duncan JS, Brooks DJ, Koepp MJ, Hammers A. Volumes, spatial extents and a probabilistic atlas of the human basal ganglia and thalamus. Neuroimage. 2007 Nov 1;**38**(2):261-270

[15] Boisgontier MP, van Ruitenbeek P, Leunissen I, Chalavi S, Sunaert S, Levin O, Swinnen SP. Nucleus accumbens and caudate atrophy predicts longer action selection times in young and old adults. Human Brain Mapping. 2016 Dec;**37**(12):4629-4639

[16] Tepper JM, Bolam JP. Functional diversity and specificity of neostriatal interneurons. Current Opinion in Neurobiology. 2004;**14**:685-692

[17] Fix JD. Basal Ganglia and the Striatal Motor System. Neuroanatomy (Board Review Series).4th ed. Baltimore, MD: Wulters Kluwer & Lippincott Williams & Wilkins; 2008. pp. 274-228

[18] Kamishina H, Yurcisin G, Corwin J, Reep R. Striatal projections from the rat lateral posterior thalamic nucleus. Brain Research. 2008:**1204**:24-39

[19] Lanciego JL, Luquin N, Obeso JA. Functional neuroanatomy of the basal ganglia. Cold Spring Harbor Perspectives in Medicine. 2012: 1;2(12):a009621. doi: 10.1101/cshperspect.a009621.

[20] Mann DM, Yates PO, Davies JS, Hawkes J. Viruses, Parkinsonism and Alzheimer's disease. Journal of Neurology Neurosurgery and Psychiatry. 1981;**44**(7):651

[21] Elizan TS, Casals J. The viral hypothesis in Parkinsonism. Journal of Neural Transmission Supplementum. 1983;**19**:75-88

[22] Kristensson K. Potential role of viruses in neurodegeneration. Molecular and Chemical Neuropathology. 1992;**16**(1-2):45-58

[23] Ringheim GE, Conant K. Neurodegenerative disease and the neuroimmune axis (Alzheimer's and Parkinson's disease, and viral infections). Journal of Neuroimmunology. 2004;**147**(1-2):43-49

[24] Harris SA, Harris EA. Herpes simplex virus type 1 and other pathogens are key causative factors in sporadic Alzheimer's disease. Journal of Alzheimers Disease. 2015;**48**(2):319-353

[25] Singhrao SK, Harding A, Poole S, Kesavalu L, Crean S. Porphyromonas gingivalis periodontal infection and its putative links with Alzheimer's disease. Mediators of Inflammation. 2015;2015:137357. doi: 10.1155/2015/137357

[26] Miklossy J, McGeer PL. Common mechanisms involved in Alzheimer's disease and type 2 diabetes: A key role of chronic bacterial infection and inflammation. Aging (Albany, NY). 2016;**8**(4):575-588

[27] Pratt BM, McPherson JM. TGF–β in the central nervous system: Potential roles in ischemic injury and neurodegenerative diseases. Cytokine & Growth Factor Reviews. 1997 Dec;**8**(4):267-292

[28] Frankola KA, Greig NH, Luo W, Tweedie D. Targeting TNF–α to elucidate and ameliorate neuroinflammation in neurodegenerative diseases. CNS & Neurological Disorders Drug Targets. 2011;**10**(3):391-403

[29] Bates MN, Fawcett J, Garrett N, Cutress T, Kjellstrom T. Health effects of dental amalgam exposure: A retrospective cohort study. International Journal of Epidemiology. 2004;**33**(4):894-902

[30] Ucar SK, Mayr JA, Feichtinger RG, Canda E, Çoker M, Wortmann SB. Previously unreported biallelic mutation in DNAJC19: Are sensorineural hearing loss and basal ganglia lesions additional features of dilated cardiomyopathy and ataxia (DCMA) syndrome? Journal of Inherited Metabolic Disease Reports. 2016:1-7

[31] Dusek P, Bahn E, Litwin T, Jabłonka–Salach K, Łuciuk A, Huelnhagen T, Madai VI, Dieringer MA, Bulska E, Knauth M, Niendorf T, Sobesky J, Paul F, Schneider SA, Czlonkowska A, Brück W, Wegner C, Wuerfel J. Brain iron accumulation in Wilson disease: A post mortem 7 Tesla MRI – histopathological study. Neuropathology and Applied Neurobiology. 2016;**66**(6):435-446

[32] Yamada K, Takahashi S, Karube F, Fujiyama F, Kobayashi K, Nishi A, Momiyama T. Neuronal circuits and physiological roles of the basal ganglia in terms of transmitters, receptors and related disorders. The Journal of Physiological Sciences. 2016;**66**(6):435-446

[33] Corwin EJ. Handbook of Pathophysiology. 3rd ed. Unit III–Integrated Control and Dysfunction–Chapter 8–Nervous System. Lippincott Williams & Wilkins; 2008

[34] Gelfand Y, Kaplitt MG. Gene therapy for psychiatric disorders. World Neurosurgery. 2013 Sep–Oct;**80**(3-4):S32 (e11–e18)

[35] Wang C. Hybrid baculovirus–adeno–associated virus vectors for prolonged transgene expression in human neural cells. Journal of Neurovirology. 2008;**14**(6):563-568

[36] Murlidharan G, Sakamoto K, Rao L, Corriher T, Wang D, Gao G, Sullivan P, Asokan A. CNS–restricted transduction and CRISPR/Cas9–mediated gene deletion with an engineered AAV vector. Molecular Therapy Nucleic Acids. 2016;**5**(7):e338

[37] Kudrna JJ, Ugen KE. Gene–based vaccines and immunotherapeutic strategies against neurodegenerative diseases: Potential utility and limitations. Human Vaccines & Immunotherapeutics. 2015;**11**(8):1921-1926

[38] Maguire–Zeiss KA, Federoff HJ. Immune–directed gene therapeutic development for Alzheimer's, prion, and Parkinson's diseases. Journal of Neuroimmune Pharmacology. 2009;**4**(3):298-308

[39] Shimamura M, Sato N, Morishita R. Experimental and clinical application of plasmid DNA in the field of central nervous diseases. Current Gene Therapy. 2011;**11**(6):491-500

[40] Abel T, Zukin RS. Epigenetic targets of HDAC inhibition in neurodegenerative and psychiatric disorders. Current Opinion in Pharmacology. 2008;**8**(1):57-64

[41] Seyhan AA. RNAi: A potential new class of therapeutic for human genetic disease. Human Genetics. 2011;**130**(5):583-605

[42] MacDonald PA, Ganjavi H, Collins DL, Evans AC, Karama S. Investigating the relation between striatal volume and IQ. Brain Imaging and Behavior. 2014 Mar;**8**(1):52-59

[43] Mühle C, Kreczi J, Rhein C, Richter–Schmidinger T, Alexopoulos P, Doerfler A, Lenz B, Kornhuber J. Additive sex–specific influence of common non–synonymous DISC1 variants on amygdala, basal ganglia, and white cortical surface area in healthy young adults. Brain Structure and Function. 2017 Mar; 222(2):881-894

[44] Reardon PK, Clasen L, Giedd JN, Blumenthal J, Lerch JP, Chakravarty MM, Raznahan A. An allometric analysis of sex and sex chromosome dosage effects on subcortical anatomy in humans. The Journal of Neuroscience. 2016 Feb 24;**36**(8):2438-2448

[45] Larson PS. Deep brain stimulation for movement disorders. Neurotherapeutics. 2014 Jul;**11**(3):465-474

[46] Franzini A, Cordella R, Penner F, Rosa M, Messina G, Rizzi M, Nardocci N, Priori A. Posteroventrolateral pallidotomy through implanted DBS electrodes monitored by recording local field potentials. British Journal of Neurosurgery. 2015;**29**(6):888-890

[47] Wichmann T, DeLong MR. Deep brain stimulation for movement disorders of basal ganglia origin: Restoring function or functionality. Neurotherapeutics. 2016 Apr;**13**(2):264-283

[48] Mavridis I, Boviatsis E, Anagnostopoulou S. Anatomy of the human nucleus accumbens: A combined morphometric study. Surgical and Radiologic Anatomy. 2011 Jul;**33**(5):405-414

[49] Mavridis I, Anagnostopoulou S. The human nucleus accumbens as a target for deep brain stimulation: Anatomic study of electrode's target point and stereotactic coordinates. Minimally Invasive Neurosurgery. 2009 Oct;**52**(5-6):212-215

[50] Benabid AL, Chabardès S, Seigneuret E, Fraix V, Krack P, Pollak P, Xia R, Wallace B, Sauter F. Surgical therapy for Parkinson's disease. Journal of Neural Transmission Supplementum. 2006;**70**:383-392

[51] Abosch A, Lozano A. Stereotactic neurosurgery for movement disorders. The Canadian Journal of Neurological Sciences. 2003 Mar;**30**(Suppl 1):S72–S82

[52] Wren J, Eriksson O, Wårdell K, Loyd D. Analysis of temperature measurement for monitoring radio–frequency brain lesioning. Medical & Biological Engineering & Computing. 2001 Mar;**39**(2):255-262

[53] Pinsker MO, Volkmann J, Falk D, Herzog J, Steigerwald F, Deuschl G, Mehdorn HM. Deep brain stimulation of the internal globus pallidus in dystonia: Target localisation under general anaesthesia. Acta Neurochirurgica (Wien). 2009 Jul;**151**(7):751-758

[54] Heida T. Electric field–induced effects on neuronal cell biology accompanying dielectrophoretic trapping. Advances in Anatomy Embryology and Cell Biology. 2003;**173**:III–IX, 1-77

[55] He F, Guan H, Zhao Z, Miao X, Zhou Q, Li L, Huang D, Liu A, Miao D. Evaluation of short–term psychological functions in opiate addicts after ablating the nucleus accumbens via stereotactic surgery. Stereotactic and Functional Neurosurgery. 2008;**86**(5):320-329

[56] Kuhn J, Lenartz D, Huff W, Mai JK, Koulousakis A, Maarouf M, Lee SH, Klosterkoetter J, Sturm V. Transient manic–like episode following bilateral deep brain stimulation of the nucleus accumbens and the internal capsule in a patient with Tourette syndrome. Neuromodulation Technology at the Neural Interface Neuromodulation. 2008;**11**(2):128-131

[57] Mavridis I, Boviatsis E, Anagnostopoulou S. Microsurgical anatomy of the nucleus accumbens: He role of the two commissures in measuring stereotactic coordinates of the target and the ablation versus the stimulation target area. Stereotactic and Functional Neurosurgery. 2010;**88**(4):264-265

[58] Aouizerate B, Cuny E, Bardinet E, Yelnik J, Martin–Guehl C, Rotge JY, Rougier A, Bioulac B, Tignol J, Mallet L, Burbaud P, Guehl D. Distinct striatal targets in treating obsessive–compulsive disorder and major depression. Journal of Neurosurgery. 2009;**111**:775-779

[59] Moran A, Bar–Gad I, Bergman H, Israel Z. Real–time refinement of subthalamic nucleus targeting using Bayesian decision–making on the root mean square measure. Movement Disorders. 2006 Sep;**21**(9):1425-1431

[60] Yılmaz Kusbecı O, Gocmen Mas N, Yuces A, Stereological evaluation of basal ganglia in Parkinson's disease. Journal of Neurological Sciences (Turkish). 2012;**29**(2):232-242

[61] Geng DY, Li YX, Zee CS. Magnetic resonance imaging–based volumetric analysis of basal ganglia nuclei and substantia nigra in patients with Parkinson's disease. Neurosurgery. 2006;**58**:256-262

[62] Ekinci N, Acer N, Akkaya A, et al. Volumetric evaluation of the relations among the cerebrum, cerebellum and brainstem in young subjects: A combination of stereology and magnetic resonance imaging. Surgical and Radiologic Anatomy. 2008;**30**:489-494

[63] Acer N, Sahin B, Usanmaz M, et al. Comparison of point counting and planimetry methods for the assessment of cerebellar volume in human using magnetic resonance imaging: A stereological study. Surgical and Radiologic Anatomy. 2008;**30**:335-339

[64] Roberts N, Puddephat M–J, McNulty V. The benefit of stereology for quantitative radiology. The British Journal of Radiology. 2000;**73**:679-697

[65] Andreasen NC, Rajarethinam R, Cizadlo T, et al. Automatic atlas–based volume estimation of human brain region from MR images. Journal of Computer Assisted Tomography. 1996;**20**:98-106

[66] Gallina P, Paganini M, Di Rita A, Lombardini L, Moretti M, Vannelli GB, Di Lorenzo N. Human fetal striatal transplantation in huntington's disease: A refinement of the stereotactic procedure. Stereotactic and Functional Neurosurgery. 2008;**86**(5):308-313

[67] Buchalla R, Hopf–Jensen S, Rubarth O, Börm W. Frameless navigated biopsy with the BrainLAB® VarioGuide system: A technical note. Journal of Neurological Surgery Part A Central European Neurosurgery. 2013 Sep;**74**(5):321-324

[68] Day R, Heilbrun MP, Koehler S, McDonald P, Peters W, Siemionow V. Three–point transformation for integration of multiple coordinate systems: Applications to tumor, functional, and fractionated radiosurgery stereotactic planning. Stereotactic and Functional Neurosurgery. 1994;**63**(1-4):76-79

[69] Kamiryo T, Laws Jr ER. An accurate adjustable applicator for magnetic resonance imaging–based stereotactic procedure using the Leksell G frame. Neurosurgery. 1999 Aug;**45**(2):397-399 (discussion 399-400)

Chapter 8

Mesencephalon; Midbrain

Ayla Kurkcuoglu

Additional information is available at the end of the chapter

http://dx.doi.org/10.5772/intechopen.68767

Abstract

The mesencephalon is the most rostral part of the brainstem and sits above the pons and is adjoined rostrally to the thalamus. It comprises two lateral halves, called the cerebral peduncles; which is again divided into an anterior part, the crus cerebri, and a posterior part, tegmentum. The tectum is lay dorsal to an oblique coronal plane which includes the aquaduct, and consist of pretectal area and the corpora quadrigemina. In transvers section, the cerebral peduncles are seen to be composed of dorsal and ventral regions separated by the substantia nigra. Tegmentum mesencephali contains red nucleus, oculomotor nucleus, thochlear nucleus, reticular nuclei, medial lemnisci, lateral lemnisci and medial longitudinal fasciculus. In tectum, the inferior colliculus and superior colliculus have main nucleus, which are continuous with the periaqueductal grey matter. The mesencephalon serves important functions in motor movement, particularly movements of the eye, and in auditory and visual processing. The mesencephalic syndrome cause tremor, spastic paresis or paralysis, opisthotonos, nystagmus and depression or coma. In addition cranial trauma, brain tumors, thiamin deficiency and inflammatory or degenerative disorders of the mesencephalon have also been associated with the midbrain syndrome.

Keywords: the midbrain, mesencephalon, crus cerebri, substantia nigra, tectum

1. Introduction

The nervous system has two components, namely the central nervous system and the peripheral nervous system. The central nervous system is composed of brain and spinal cord. The peripheral nervous system consists of sensory neurons, ganglia and nerves connecting with each other and with the central nervous system. The brain is a component of the central nervous system. It contains three basic subdivisions, namely the cerebral hemispheres, brainstem and cerebellum (**Figure 1**) [1, 2].

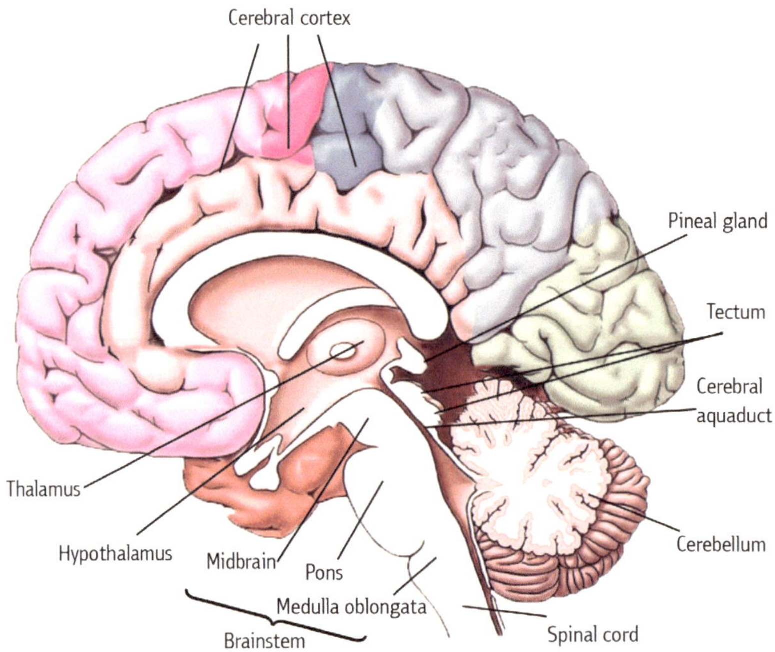

Figure 1. Mid-sagittal section of the brain.

2. Embryology

Embryologically, the central nervous system can be divided into five continuous parts (**Figure 1**). From rostral to caudal, they are [3, 4]:

- The cerebrum (telencephalon) becomes two cerebral hemispheres. The surface of these hemispheres consists of gyri and sulci, and the hemispheres are partially separated by a deep longitudinal fissure.
- The diencephalon is hidden from view in the adult brain by the cerebral hemispheres. It consists of the thalamus, hypothalamus, and other related structures and classically is considered to be the most rostral part of the brainstem.
- The mesencephalon (midbrain), which is the first part of the brainstem seen when an intact adult brain is examined, located at the junction between and in both middle and posterior cranial fossae.
- The metencephalon gives rise to the cerebellum and the pons.
- The myelencephalon (medulla oblongata), the caudal most part of the brainstem, ends at the foramen magnum.

Closure of the neural tube first occurs in the region where the earliest somites appear; closure spreads cranially and caudally. The unfused regions of the neural tube are known

http://dx.doi.org/10.5772/intechopen.68767

as the cranial and caudal neuropores. Even before the closure of the neuropores (24 days of gestation for the cranial neuropore, and 26 days of gestation for the caudal neuropore), some fundamental subdivisions in the early nervous system have become manifest. The future spinal cord and brain are recognizable, and within the brain the forebrain (prosencephalon), midbrain (mesencephalon), and hindbrain (rhombencephalon) can be distinguished [4]. A prominent force in shaping the early nervous system is the overall bending of the cephalic end of the embryo into a "C" shape. Associated with this bending is the appearance of a prominent cephalic flexure of the brain at the level of the mesencephalon at the end of the third week. At the beginning of the fifth week, a cervical flexure appears at the boundary between the hindbrain and the spinal cord. By week 5, the original three-part brain has become subdivided further into five parts. The mesencephalon, which is sharply bent by the cephalic flexure, remains undivided and tubular in its overall structure [2, 4].

3. Anatomy

3.1. External features of the midbrain

The midbrain is the short, constricted portion, which connects the pons and cerebellum with the thalamus and cerebral hemispheres. It is the smallest part of the brainstem, not more than 2 cm in length, and most of it lies in the posterior cranial fossa [3, 5].

On the anterior surface of the midbrain are located the cerebral peduncles, separated by the interpeduncular fossa. The cerebral peduncle is two large bundles of fibers on each side of the midline. In addition, it is again divided into an anterior part, *the crus cerebri,* and a posterior part, *tegmentum mesencephali,* by a pigmented band of gray matter called *substantia nigra. The crura cerebri* are superficially corrugated and emerge from the cerebral hemispheres. They converge as they descend and meet as they enter the pons, where they form the caudolateral boundaries of the interpeduncular fossa. Two crura are separate, whereas the tegmental parts are united and traversed by the cerebral aqueduct that connects the third and fourth ventricles [1, 2]. The oculomotor nerve (CN III) emerges from the medial aspect of the cerebral peduncle of the same side. Exiting from the interpeduncular fossa near the junction of the pons and midbrain are the oculomotor nerves. This cranial nerve supplies all but two of the extraocular muscles. The crus cerebri embraces a midline depression called the interpeduncular fossa. The basilar artery divides in the interpeduncular fossa into right and left posterior cerebral arteries at the level of the tentorial incisura. The superior cerebellar and posterior cerebral arteries run laterally around the ventral (basilar) crural surfaces. The trochlear and oculomotor nerves lie between two arteries. In the depths of the interpeduncular fossa can be seen numerous small holes. These holes represent the entry point of the posterior cerebral artery. Because of its appearance, this region is usually referred to as the posterior perforated substance [6–8].

On the posterior surface of the midbrain are four prominent rounded elevations, the inferior and superior colliculi one on each side. Collectively, four colliculi are called as *tectum* (corpora quadrigemina). The superior and inferior colliculi are separated by a cruciform sulcus. The upper limit of the sulcus expands into a depression for the pineal gland. Median frenulum veli is prolonged from its caudal end down over the superior medullary velum. The superior colliculi are larger and darker than the inferior colliculi, and associated with visual responses. The inferior colliculi are smaller, and associated with auditory pathways. Each colliculus is laterally related to ridges called superior brachium and inferior brachium, coming from respective colliculi. Superior brachium connects the superior colliculus to lateral geniculate body. Inferior brachium connects the inferior colliculus to medial geniculate body. The trochlear nerves (CN IV) arise from the dorsal midbrain, caudal to the inferior colliculi and pass inferiorly around the lateral side of the midbrain. The trochlear nerve is the only cranial nerve that exits from the dorsal surface of the brainstem. The midbrain serves important functions in motor movement, particularly movements of the eye, and in auditory and visual processing [1, 2, 7].

3.2. Internal structure of the midbrain

On transverse section, the cerebral peduncles are seen to be composed of dorsal and ventral regions separated by *substantia nigra*. On each side, the dorsal region is *tegmentum,* and the ventral part is the *crus cerebri*. Cerebral peduncles are the major pathways of motor neurons out of the cortex. *The tegmentum* is between the *substantia nigra* and the *aquaductus mesencephali*. It also refers to the corresponding regions in the medulla and pons. *Tectummesencephali,* located dorsal to the *aquaductus mesencephali,* contains two superior colliculi and two inferior colliculi (**Figures 2** and **3**) [1, 2, 6].

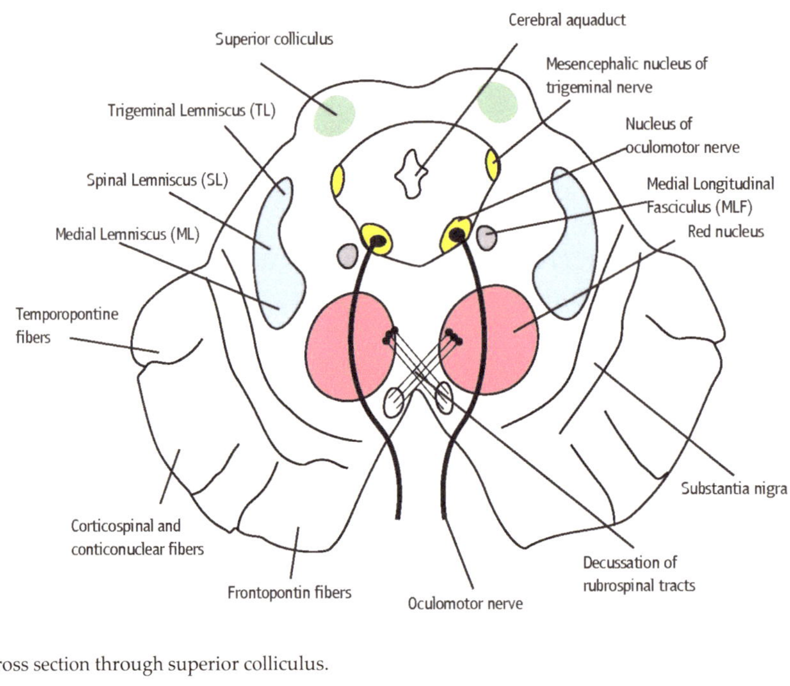

Figure 2. Cross section through superior colliculus.

http://dx.doi.org/10.5772/intechopen.68767

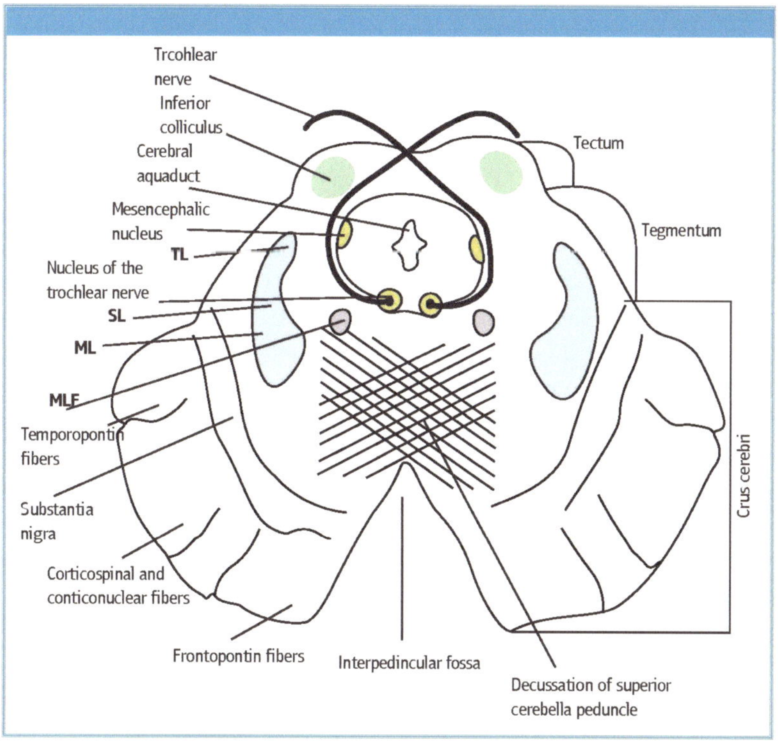

Figure 3. Cross section through inferior colliculus.

3.2.1. Crura cerebri (the cerebral peduncles, pedunculus cerebri)

The most ventral part of the midbrain contains a massive band of descending corticofugal fibers, the crus cerebri. Each crus cerebri is semilunar in section. It contains corticonuclear, corticospinal and corticopontine fibres. Corticonuclear fibers (the corticobulbar fibers) originate in the primary motor cortex. They descend through the genu of the internal capsule, and down to the midbrain. In the midbrain, the middle third of the crus cerebri contains the corticobulbar and corticospinal fibers. The corticobulbar fibers end in the motor nuclei of the cranial nerves and other brainstem nuclei. The corticospinal fibers are white matter motor pathways starting at the cortex, and they travel through the posterior limb of the internal capsule. They enter the cerebral peduncle at the base of the midbrain, then pass through the brainstem, from the pons and then to the medulla. Corticospinal neurons synapse directly onto alpha motor neurons in the spinal cord for direct muscle control. Corticopontine fibers arise in the cerebral cortex and form two groups, both of which end in pontine nuclei. The frontopontine fibers arise from the cells of the frontal lobe, and end in the nuclei of the pons, the temporopontine fibers, which are largely from the posterior region of the temporal lobe, traverse the internal capsule, but occupy the lateral sixth of the ipsilateral crus [1, 5, 7].

3.2.1.1. Substantia nigra

The pigmented substantia nigra, the largest single nuclear mass in the midbrain, is connected massively with the basal ganglia, but it has other projections as well it is considered to subserve a motor function. It looks like a darkened streak in unstained brain tissue; this is where it gets its name, which is Latin for "black substance." Although it is often referred to as one structure, the substantia nigra is actually made up of two anatomically and functionally distinct portions: the substantia nigra pars compacta and the substantia nigra pars reticulata [2, 3]. Neurons in the pars compacta are much more densely packed together (or compact) than those in the pars reticulata. The compact zone appears as an irregular band of closely packed, large polygonal or pyramidal cells containing granules of melanin pigment. The reticular zone, also known as the stratum intermedium, lies close to the crus cerebri, and is composed of irregular shaped scattered cells that are rich in iron, but they do not contain melanin pigment. Most of the dopamine neurons of the brain originate in the midbrain and are found in either the substantia nigra or the ventral tegmental area, which is located adjacent to the substantia nigra [6, 8]. These dopamine neurons, however, are found predominantly in the substantia nigra pars compacta. The pars reticulata is instead populated largely by GABA neurons. Lesions of the substantia nigra or dopamine deficiencies result in Parkinson's disease. Although it is still not clear what exactly causes neurodegeneration in Parkinson's disease, when a significant number of these neurons have died, the individual will likely start to experience movement-related problems such as tremor, rigidity, slowing of movements, and postural instability—all hallmark symptoms of Parkinson's disease [9].

Afferent fibers to the substantia nigra arise mainly from the caudate nucleus and the putamen. The efferent fibers of the substantia nigra project to the striatum and certain thalamic nuclei.

3.2.2. Mesencephalic tegmentum

Mesencephalic tegmentum is between the substantia nigra and the cerebral aqueduct. It usually contains ascending fiber tracts, cranial nerve nuclei, and the reticular formation nuclei.

The structures in this section are listed below:

3.2.2.1. Red nucleus

The red nucleus is paired, oval shaped, and approximately 5 mm in diameter, midline structure that appears red in a freshly dissected specimen. This unique appearance has been attributed to high vascularity of the structure in addition to the high level of iron pigments in the cytoplasms of its constituent neurons. It blends rostrally with the nearby reticular formation and interstitial nucleus. The nucleus consists of a large and ovoid column of cells extending from the caudal margin of the superior colliculus into the caudal diencephalon, and it appears as a circular mass which is traversed by the fibers of the oculomotor nerve [1, 3, 7].

Afferent fibers projecting to the red nucleus are derived from two principal sources, the cerebellar nuclei (approximately half of the fibers arising from the dentate nucleus pass rostrally beyond the red nucleus, cerebellorubral fibers), and the cerebral cortex (mainly from

http://dx.doi.org/10.5772/intechopen.68767

the precentral gyrus, corticorubral fibers). Fibers of the superior cerebellar peduncle, arising from the dental, globose and emboliform nuclei, undergo a complete decussation in the caudal midbrain, and both enter and surround the contralateral red nucleus.

The efferent fibers of the red nucleus continue through the spinal cord, the brainstem and the cerebellum. Rubrospinal fibers issue from the medial margin of the red nucleus. The fibers decussate and then run obliquely laterally in the ventral tegmental decussation (of Forel), ventral to the tectospinal decussation and dorsal to the medial lemniscus. Some efferent axons from a rubrobulbar tract to motor nuclei of the trigeminal, facial, oculomotor, trochlear and abducens nerves. Uncrossed descending rubral efferents, from the parvocellular part of the nucleus, enter the central tegmental tract and project to the dorsal lamella of the principal inferior olivary nucleus. These fibers are referred to as rubro-olivary fibers Physiology [3, 6, 7].

In humans, the lesions of a rubrospinal system are described as producing a syndrome characterized by contralateral motor disturbances that are variously designated as tremor, ataxia and choreiform activity and ibsilateral oculomotor palsy [1, 10].

3.2.2.2. *Decussation of the superior cerebellar peduncle*

All fibers of the superior cerebellar peduncle decussate at levels through the inferior colliculus. The decussation of superior cerebellar peduncle is the crossing of fibers of the superior cerebellar peduncle across the midline. It comprises the cerebellothalamic tract, which arises from the dentate nucleus, as well as the cerebellorubral tract, which arises from the globose and emboliform nuclei and project to the contralateral red nucleus to eventually become the rubrospinal tract [2, 7, 8].

It is important as an anatomical landmark, as lesions above it cause contralateral cerebellar signs, while lesions below it cause ipsilateral cerebellar signs.

3.2.2.3. *Trochlear nucleus*

The nucleus of the trochlear nerve is small and nearly circular and is on a level with a plane carried transversely through the upper part of the inferior colliculus. Root fibers emerging from the nucleus curve dorsolaterally and caudally in the outer margin of the central gray decusate completely in the superior medullary velum and from the dorsal surface of the brainstem caudal to the inferior colliculus [1, 2]. Root fibers emerging from the nucleus curve dorsolaterally and caudally in the outer margin of the central gray decussate completely in the superior medullary velum and exit from the dorsal surface of the brainstem caudal to the inferior colliculus. And then, it curves around the lateral surface of the brainstem, passes between the superior cerebellar and posterior cerebral arteries, and enters the cavernous sinus. It innervates the superior oblique muscle that serves to: intort the eye when abducted, and depress the eye when adducted. The trochlear nerve (CN IV) is the smallest nerve in terms of the number of axons it contains, and it has the longest length intracranially [1, 3, 5]. There are two major clinical syndromes that can manifest through damage to the trochlear nerve:

1. Vertical diplopia: Injury to the trochlear nerve causes weakness of downward eye movement with consequent vertical diplopia.
2. Torsional diplopia: Weakness of intorsion results in torsional diplopia, in which two different visual fields, tilted with respect to each other, are seen at the same time. To compensate, patients with trochlear nerve palsies tilt their heads to the opposite side in order to fuse the two images into a single visual field.

The clinical syndromes may originate from both peripheral and central lesions. A peripheral lesion is damage to the bundle of nerves, in contrast to a central lesion, which is damage to the trochlear nucleus [8, 10].

A lesion of the trochlear nucleus affects the contralateral eye. Lesions of all other cranial nuclei affect the ipsilateral side [10].

3.2.2.4. Parabigeminal nucleus

Ventrolateral to the inferior colliculus is a fairly well-defined zone known as the parabigeminal area. It is between the lateral lemniscus and the inferior colliculus and the surface of the brainstem that contains the parabigeminal nucleus. Some of the superior collicular efferents pass to the parabigeminal nucleus. Studies have shown that the parabigeminal nucleus is active with fixed or moving objects. In addition, the parabigeminal nucleus plays a role in assessing the vision together with the colliculus superior [2, 3, 7].

3.2.2.5. Posterior commissure

The posterior commissure is located in the inferior pineal lamina and lies rostral to the superior colliculus at the place where the cerebral aqueduct becomes the third ventricle. It is one of the commissural fibers of the brain known to be important in the pupillary light reflex. Various nuclei are associated with the posterior commissure. The best known of them is the interstitial nucleus of the posterior commissure, nucleus of Darkschewitsch; another one is the interstitial nucleus of Cajal. Fibers from the olivary nucleus cross on the opposite side and give collaterals to the visceral nuclei of the oculomotor complex (Edinger-Westphal nucleus) [2, 3, 11]. Some fibers are believed to be derived from the posterior part of the thalamus and from the superior colliculus and to continue directly to the medial longitudinal fasciculus. Fibers from the thalamic, pretectal, tectal region, and the habenular nuclei are known to connect with the posterior commissure, but they have not been shown anatomically.

Lesions in the nuclei of the posterior commissure, interrupting fibers from the intersititial nuclei of Cajal, produce bilateral eyelid retraction and impairment of vertical eye movements [12].

3.2.2.6. Pretectal area and pretectal nucleus

The pretectal region lies rostral to the superior colliculus at levels of the posterior commissure and rostrally by the habenular trigone, and laterally by the pulvinar thalami. This area is composed of several distinct cell groups, most of which are related to the visual system. The nuclei of the pretectal area include pretectal olivar nucleus, medial, anterior and posterior pretectal nuclei, and optic tract nucleus [2, 13].

Afferents come from the lateral root of the optic tract from the retina, occipital cortical fields via the superior quadrigeminal brachium and the superior colliculus. Efferents go to the ipsilateral and contralateral accessory oculomotor nucleus and superior colliculus. They have bilateral efferent connections with the Edinger-Westphal nucleus of the oculomotor nuclear complex by way of which they mediate the pupillary light reflex. The effrent fibers reach both Edinger-Westphal nucleui. The decussate fibers pass ventral to the aqueduct or through the posterior commissure. In this way, sphincter pupillae contract in both eyes in response to impulses from either eye. Therefore, only relatively large lesions involving multiple structures in the pretectum appear to impair the pupillary light reflex [2, 8, 13].

3.2.2.7. Medial longitudinal fasciculus

The medial longitudinal fasciculus is situated in the brainstem, and it is a set of crossed fibers with ascending and descending fibers. The medial longitudinal fasciculus is a heavily myelinated composite tract lying near the midline, ventral to the periaqueductal gray matter. It links three main nerves which control eye movements including saccades (rapid refixations), that is, the oculomotor, trochlear and the abducent nerves, as well as the vestibulocochlear nerve. It interconnects the oculomotor, troclear, abducens, Edinger-Westphal, vestibular, reticular and spinal accessory nuclei, coordinating conjugate eye movements and associated movements of the neck and head, including semicircular- and otolith-mediated ocular motor reflexes [1, 12, 14]. The medial longitudinal fasciculus provides a neural mechanism for simultaneous contraction of the lateral rectus muscle on the one side, and the medial rectus muscle on the opposite side, required for conjugate lateral gaze. It also forms a major component of the optokinetic and vestibule-ocular reflexes.

The most commonly recognized syndrome that results from the medial longitudinal fasciculus damage is internuclear ophthalmoparesis, which is characterized by slowing or limitation of adduction (on the same side as the medial longitudinal fasciculus lesion) during horizontal eye movements. In patients with internuclear ophthalmoparesis, the contralateral abducting eye will usually exhibit a disassociated horizontal nystagmus, although this does not always occur [13, 15].

3.2.2.8. Oculomotor nucleus

The oculomotor nucleus is on the level of colliculus superior. The upper end of the nucleus approaches the bottom of third ventricle in the vicinity of commissura caudalis and the lower end the top of colliculus inferior, lying adjacent to the inner ventral edge of stratum griseum centrale surrounding aquaeductus mesencephali. The nucleus of the oculomotor nerve is about 10-mm long. The oculomotor nuclear complex containing subnuclei that give rise to the axons of the occulomotor nerve, both motor and parasympathetic fibers, is situated at the midline, at the level of the superior colliculus in the midbrain tegmentum [8]. The complex including the somatic portion is formed by multipolar motor neurons, and the parasympathetic portion is formed by oval or fusiform preganglionic cells, on each side of the median raphe. The somatic portion consists of the lateral somatic cell column and the caudal central nucleus. It is divided into the principal, intrafascicular and extrafascicular parts. The principal part is subdivided into the dorsal, intermediate and ventral portions. Isolated multipolar neurons are also found in the periaqueductal gray matter, the interstitial nucleus of Cajal, the Edinger-Westphal nucleus, and the fiber bundles of the oculomotor nerve. These cells most likely represent the displaced motor neurons of the oculomotor nerve. The Edinger-Westphal nucleus consists of the rostral, ventral and dorsal parts [16]. The lateral somatic cell columns innervate the extraocular muscles. The dorsal column innervates the inferior rectus muscle. The intermediate cell column innervates the inferior ollique muscle, and the ventral cell column supplies fibers to the madial rectus muscle. A cell column medial to both dorsal and intermediate cell column is referred to as the medial cell column, and it innervates the superior rectus muscle. The caudal central oculomotor nucleus is a midline somatic cell group found only in the caudal third of the complex, and it innervates the levator palpebrae muscle. As a result, the oculomotor nuclear complex innervates all extraocular muscles except the lateral rectus and the superior oblique, supplies the levator palpebrae muscle, and provides preganglionic parasympathic fibers to the ciliary ganglion [17]. Visceral nuclei of the oculomotor nuclear complex consist of two distinct nuclear groups which are in continuity rostrally, and often are collectively referred to as the Edinger-Westphal nucleus. Its afferent inputs come from the pretectal nuclei bilaterally and mediate the pupillary light reflex. In addition, they come from the visual cortex, mediating accommodation. Efferent fibers relay through the ciliary ganglion and synapse upon postganglionic neurons, which give rise to the short ciliary nerves. The postganglionic fibers innervate the ciliary body, concerned with the mechanism of accommodation, and the sphincter of the iris [1, 17, 18].

Oculomotor nerve palsy results in weakness of the medial rectus, inferior rectus, superior rectus, inferior oblique, and levator palpebrae, leading to an eye that is "out and down ". The oculomotor nerve also carries parasympathetic innervation to the pupil, responsible for pupil constriction. Oculomotor palsy may therefore leave the pupil dilated. Nuclear lesions usually occur due to small regions of infarction, and often there are no other neurological symptoms. In ventral midbrain, lesions are due to HYPERLINK "https://radiopaedia.org/articles/benedikt-syndrome" Benedikt syndrome and HYPERLINK "https://radiopaedia.org/articles/weber-syndrome" Weber syndrome [2, 5, 19, 20].

Pupillary light reflex consists of a simultaneous and equal constriction of the pupils in response to illumination of one or the other eye. The afferent axons of retinal ganglion cells pass into the optic nerve and decussate in the chiasm, and pass with the optic tract to the midbrain. The pupillary fibers do not synapse with the visual fibers in the lateral geniculate body, but pass to the pretectal nuclei at the level of the superior colliculus with intercalated fibers that pass as the efferent pupillary pathway to the Edinger-Westphal nucleus of the oculomotor nerve on both sides. Preganglionic parasympathetic fibers run in the oculomotor nerve as it leaves the brainstem. The fibers pass downward to lie inferiorly in the inferior division of the third nerve as it enters the orbit. These fibers synapse in the ciliary ganglion and give rise to postganglionic parasympathetic myelinated short ciliary nerves [1,19, 20, 21].

The accommodation-convergence reaction occurs when gaze is shifted from a distant object to a near one. The accommodation reflex has its afferent input from the primary visual pathway; sequentially retina, optic nerve, optic chiasm, optic tract, lateral geniculate body, optic radiation, visual cortex and area 17. The peristriate area 19 interprets accommodation, and sends signals via the Edinger-Westphal nucleus and the oculomotor nerve to the ciliary muscle, the medial rectus muscle and the sphincter pupillae muscle [20, 21].

3.2.2.9. Mesencephalic reticular formation

The mesencephalic reticular formation is less extensive than the pontine reticular formation caudal to it. It is a neuronal structure located in the core of the brainstem, its caudal boundary crosses of the superior cerebellar peduncle, and extends rostrally to the thalamic reticular nucleus. It is reciprocally interconnected with the superior colliculus. Even though detailed studies have indicated that the red nucleus is recognized as a distinctive part of the reticular formation, the principal reticular nuclei of the mesencephalon are: 1. The pedinculopontine nucleus, 2. The nucleus cuneiformis, and 3. The nucleus subcuneiformis. Original function is defined as a part of the reticular activating system. There is clear evidence showing that there are subgroups of cells that participate in the control of saccadic and vergence eye movements. The mesencephalic reticular formation has two major subdivisions. The posterior commissure in the sub-human primate serves to separate the mesencephalic reticular formation into rostral and caudal regions. The cells of the rostral portion of the mesencephalic reticular formation are associated with the control of vertical eye movements, while neurons in the caudal region also called as the central mesencephalic reticular formation are more closely associated with the control of horizontal eye movements [1, 7, 22].

3.2.2.10. Interpedincular nuclei

It is a ventral nucleus of the midbrain tegmentum lying between the right and left substantia nigrae, which cap the two cerebral peduncles. The interpeduncular nucleus receives axons from the habenula, and it sends axons dorsally, to the midbrain raphe nuclei. The pathways described above constitute a part of the complex system by which impulses related to the limbic system are projected to midbrain levels [1–4].

3.2.2.11. The mesencephalic nucleus of the trigeminal nerve

The mesencephalic nucleus of the trigeminal nerve that is composed of large unipolar neurons forms a slight cell column near the lateral margin of the central gray of the upper part of the cerebral aqueduct. The nucleus extends from the level of the motor nucleus into the rostral midbrain. The cell bodies located in mesencephalic nucleus of the trigeminal nerve are actually connected to primary sensory fibers primarily coming from masticatory muscles. This is the only place in the nervous system where the cell bodies of primary afferent fibers are found in the central nervous system rather than in ganglia outside of it. It houses proprioception for all muscles of the head and face. This nucleus has connections to the motor nucleus of the trigeminal nerve [1, 3, 13].

Afferent fibers of the mesencephalic nucleus of the trigeminal nerve convey proprioceptive impulses (kinesthesis and pressure) from the teeth, periodontium, hard palate, muscles of mastication and joint capsules. The mesencephalic nucleus of the trigeminal nerve receives afferent impulses from stretch receptors in the muscles of mastication. The jaw proprioception pathway consists of sensory pseudounipolar neurons. Their peripheral processes consist of stretch receptors that terminate in the muscles of mastication. Their central processes bifurcate to send a branch to the principal sensory nucleus, and another branch to the rostral portions of the spinal nucleus. Second-order neurons from these nuclei project to the ventral posteromedial nucleus of the thalamus, which in turn projects to Brodmann's area 3 a on the medial surface of the primary somatosensory cortex in the parietal lobe where position sense of the mandible enters conscious awareness. Although proprioceptive information is generally processed by the mesencephalic neurons, their receptors, and their central connections, some proprioception is relayed from the temporomandibular joint, and extraocular muscle spindles [2–5, 13].

3.2.2.12. Sensory tracts of the tegmentum

Medial lemniscus*:* It is formed by the crossings of internal arcuate fibers, composed of axons of nucleus cuneatus and nucleus gracilis. It ascends from the lemniscal decusstio on each side, as a flattened tract near the median raphe. This large ascending fiber bundle can be readily followed through the brainstem to its termination in the ventral posterolateral nucleus of the thalamus. On the transverse section of the midbrain, the medial lemniscus is dorsal in tegmentum. Medial lemniscus is important for somatosensation from the joints and skin, and therefore, lesions of the medial lemnisci cause an impairment of vibratory and touch-pressure sense [1, 2, 5].

Lateral lemniscus: The lateral lemniscus projects to the contralateral side from the dorsal cochlear nucleus. The ventral cochlear nuclei project to the superior olivary nuclei on both sides as well as into the lateral lemniscus on each side. Thus, the lateral lemniscus contains axons originating from cells in the dorsal and ventral cochlear nuclei and in the superior olivary nucleus. The lateral lemniscus terminates in the central nucleus of the inferior colliculus. The principal ascending auditory pathway in the brainstem courses rostrally in the lateral part of the tegmentum [3, 6, 13].

3.2.3. Tectum

The tectum is located in the dorsal part of the midbrain, below the diencephalon. The name comes from the Latin word for "roof." Tectum is composed of a set of *colliculi superior* and *colliculi inferior,* which resemble small lumps and are responsible for initial processing of sensory information from the eyes and ears. Tectum is responsible for visual and auditory reflexes [3, 5, 13].

3.2.3.1. Superior colliculi

The superior colliculus refers to the rostral bump on the lateral side of the midbrain. It receives afferents from a number of sources including the retina, spinal cord, inferior colliculus and occipital and temporal cortices. Collicular efferents pass to retina, lateral geniculate nucleus, pretectum, parabigeminal nucleus, thalamus and spinal cord. They pass through the pulvinar relay to primary and secondary visual cortices. The tectobulbar and tectospinal tracts start from neurons in the superior colliculi and sweep ventrally round the central gray matter to decussate ventral to the oculomotor nuclei and medial longitudinal fasciculi as part of the dorsal tegmental decussations (of Meynert) (**Figure 2**). The superior colliculus is not restricted to a visual role alone. It also helps orientation of the eyes and head. Part of the colliculus sticks out in the direction of the spinal cord region. This key projection helps the head to respond to different sensory stimuli. The superior colliculus is concerned primarily with the detection of the direction of movement of object in the visual fields, and in this way, it facilitates visual orientation, searching and tracking [1, 2, 8].

The collicular stimulation produces contralateral head movement as well as movements involving the eyes, trunk and limbs, which implicates the superior colliculus in complex integration between vision and widespread body activity.

3.2.3.2. Inferior colliculi

The inferior colliculus is a part of the midbrain that serves as a main auditory (sound) center for the body (**Figure 3**). It consists of a compact nucleus of gray substance containing large and small multipolar nerve cells, and more or less completely surrounded by white fibers derived from the lateral lemniscus. Its primary roles are signal integration, frequency recognition, and pitch discrimination. It also processes sensory signals from the superior colliculi, located above it. The inferior colliculi is the relay station for Auditory Pathway. It receives fibers from the lateral lemniscus, the opposite inferior colliculus, the ipsilateral medial geniculate body and the auditory cortex. Most efferent fibers travel via the inferior brachium to the ipsilateral medial geniculate body. Some colliculogeniculate fibers do not relay in the geniculate body, but continue, with those that do, via the auditory radiation to the auditory cortex area. In few, they pass to the opposite inferior colliculus and the superior colliculus [2, 5, 8].

In experimental animals, lesions of either the inferior colliculus or its brachium produce defects in tonal discrimination, sound localization and auditory reflex. The effects of such lesions are poorly documented in humans (**Table 1**) [23].

4. Blood supply of the midbrain

The brain receives blood from two sources: the internal carotid arteries and the vertebral arteries. The internal carotid arteries arise at the point in the neck where the common carotid arteries bifurcate. It branches to form two major cerebral arteries (terminal branches): the anterior and middle cerebral arteries. The vertebral arteries (right and left) arise from the subclavian artery. They come together on the ventral surface of the brainstem at the level of the pons to form the midline basilar artery. The basilar artery joins the blood supply from the internal carotids in an arterial ring called the circle of Willis. Conjoining two major sources of cerebral vascular supply via the circle of Willis presumably improves the chances of any region of the brain continuing to receive blood if one of the major arteries becomes occluded. The basilar artery undergoes bifurcation at the site of midbrain, forming two posterior cerebral arteries. Both posterior cerebral arteries travel around the cerebral peduncles, and branch into the midbrain forming a series of slender, long penetrating arteries that are responsible for supplying blood to the thalamus and hypothalamus [2, 3, 8, 24].

The mesencephalon receives its blood supply principally from branches of the basilar artery, although branches of the internal carotid also contribute. The main vessels supplying this portion of the brainstem include:

1. The posterior cerebral artery (terminal branch of the basilar artery)
2. The superior cerebellar artery (branch of the basilar artery)
3. Branches of the posterior communicating artery (branch of the internal carotid artery)
4. Branches of the anterior choroidal artery (branch of the internal carotid artery)

Numerous veins of the mesencephalon arise from capillaries and, in general, run near the arteries. These veins from an extensive peripheral plexus in the pia and are collected by the basal veins which drain into either the great cerebral vein (Galen) or the internal cerebral veins [1, 2, 8].

5. Lesions of the midbrain (midbrain lesions)

Lesions of the midbrain are described in the following (**Table 1**) [2, 13, 24, 25].

	Lesions	Symptoms
Parinaud's syndrome (dorsal midbrain syndrome)	It localizes pathology to impingement of or origin in the tectal plate, most frequently due to a posterior commissure or pineal region mass (typically solid tumors rather than pineal cysts) (**Figure 4**)	Convergence-retraction nystagmus pupillary light-near dissociation upward gaze palsy, often manifesting as diplopia
Benedict syndrome (paramedian midbrain syndrome)	It is a rare form of posterior circulation stroke of the brain,	Oculomotor nerve (CN III) palsy
	A lesion within the tegmentum of the midbrain can produce this syndrome (**Figure 5**)	Cerebellar ataxia including tremor Neuroanatomical structures affected include: red nucleus, corticospinal tracts, brachium conjunctivum, the superior cerebellar peduncle decussation
Weber's syndrome	Paramedian infact of midbrain,	The lesioned substantia nigra causes contralateral parkinsonism

http://dx.doi.org/10.5772/intechopen.68767

	Lesions	Symptoms
(Superior alternating hemiplegia)	A lesion within the crus cerebri of the midbrain can produce this syndrome (**Figure 6**)	Damage to the corticobulbar tract will produce difficulty with contralateral lower facial muscles and hypoglossal nerve functions
		Damage to the corticospinal tract will produce contralateral hemiparesis and typical upper motor neuron findings
		Damage to the oculomotor nerve fibers lead to ipsilateral oculomotor nerve palsy with a drooping eyelid and fixed wide pupil pointed down and out this leads to diplopia

Table 1. Lesions of the midbrain.

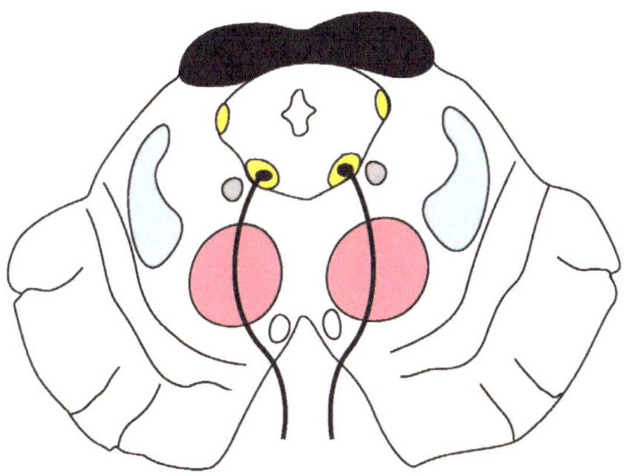

Figure 4. Parinaud's syndrome (dorsal midbrain syndrome).

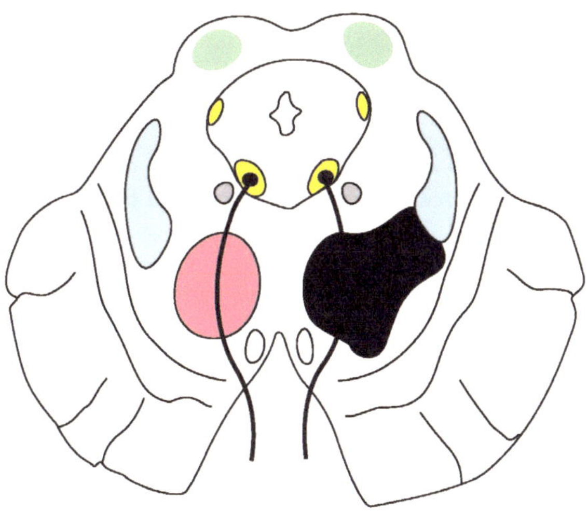

Figure 5. Benedict syndrome (paramedian midbrain syndrome).

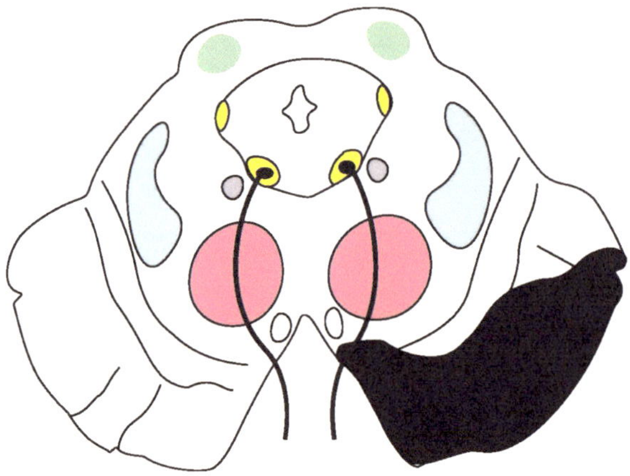

Figure 6. Weber's syndrome (superior alternating hemiplegia).

Author details

Ayla Kurkcuoglu

Address all correspondence to: kayla@baskent.edu.tr; kurkcuoglua@hotmail.com.tr

Department of Anatomy, Faculty of Medicine, Baskent University, Baglıca, Ankara, Turkey

References

[1] Carpenter MB. Human Neuroanatomy. 7th ed. Baltimore, USA: Williams & Wilkins; 1976. Pp. 657-662

[2] Moore KL, Dalley AF. Clinical Oriented Anatomy. 5th ed. Philadelphia: Lippincott Williams and Wilkins; 2002. pp. 929-935

[3] Gray's Anatomy. The Anatomical Basis of Clinical Practice. 39th ed. USA: Elsevier Crushill Livingstone; 2005. pp. 340-349

[4] Carlson BM. Human Embryology and Developmental Biology. 4th ed. Mosby Elsevier; 2009. Pp. 239-245

[5] Martin HJ. Neuroanatomy Text and Atlas. 2nd ed. 1996. pp. 522-525

[6] Brodal P. The Central Nervous System: Structure and Function. New York: Oxford University Press; 2004

[7] Patestas MA, Gartner LP. A Textbook of Neuroanatomy. John Wiley & Sons; 2016. p. 328

[8] Brodal P. Neurological Anatomy. 3rd ed. New York; 1981. pp. 533-534

[9] http://www.neuroscientificallychallenged.com/blog/know-your-brain-substantia-nigra

[10] https://www.boundless.com/physiology/textbooks/boundless-anatomy-and-physiology-textbook/peripheral-nervous-system-13/cranial-nerves-131/trochlear-iv-nerve-701-8443/

[11] Ozdemir NG. The anatomy of the posterior commissure. Turk Neurosurg. 2015;**25**(6):837-43. DOI: 10.5137/1019-5149.JTN.12122-14.2

[12] Carpenter MB, Peter P. Accessory oculomotor nuclei in the monkey. The Journal für Hirnforschung. 1970;**12**:405-418

[13] Snell RS. Clinical Anatomy by Systems. Baltimore: Lippincott Williams and Wilkins; 2007. pp. 557-570

[14] Paul R. Spinal Tracts—Descending/Motor Pathways Essential Clinical Anatomy of the Nervous System, Chapter 9. 2015. pp 161-176. DOI: http://dx.doi.org/10.1016/B978-0-12-802030-2.00009-1

[15] Frohman TC, Galetta S, Fox R, Solomon D, Straumann D, Filippi M, Zee D, Frohman EM. Pearls & Oy-sters. The medial longitudinal fasciculus in ocular motor physiology. Neurology. 2008, 22;**70**(17):e57-67. DOI: 10.1212/01.wnl.0000310640.37810.b3

[16] Donzelli R, Marinkovic S, Brigante L, Nikodijevic I, Maiuri F, de Divitiis O. The oculomotor nuclear complex in humans. Microanatomy and clinical significance. Surgical and Radiologic Anatomy. 1998;**20**(1):7-12

[17] Saiyo A. Microscopical study of oculomotor nucleus in human adult. The Tohoku Journal of Experimental Medicine. 1953;**5**:3-4

[18] Purves D, Augustine GJ, Fitzpatrick D. Neuroscience. 2nd ed. Sunderland (MA): Sinauer Associates; 2001

[19] Kaufman, Paul L, Levin, Leonard A, Alm, Albert. Adler's Physiology of the Eye. Elsevier Health Sciences; 2011. p. 508

[20] Lo CP, Huang CF, Hsu CC. Neuroimaging of isolated and non-isolated third nerve palsies. The British Journal of Radiology. 2012;**85** (1012):460-467. DOI: 10.1259/bjr/38090653AkdalG

[21] Waugh A, Grant A. Anatomy and Physiology in Health and Illness. 11th ed. Toronto: Churchill Livingstone Elsevier; 2010. pp. 147-154

[22] Binder DM, Hirokawa N, Windhorst U. Mesencephalic Reticular Formation (MRF). Encyclopedia of Neuroscience. 2009. pp. 2339-2339. DOI: 10.1007/978-3-540-29678-2_3444

[23] http://www.healthline.com/human-body-maps/inferior-colliculus

[24] Ruchalski K, Hathout GM. A medley of midbrain maladies: A brief review of midbrain anatomy and syndromology for radiologists. Radiology Research and Practice. Volume 2012, Article ID 258524, 11 pages. DOI: 10.1155/2012/258524

[25] Kutluk K, Men S, Yaka E. Benedikt and "plus-minus lid" syndromes arising from posterior cerebral artery branch occlusion. Journal of the Neurological Sciences. 2004;**228**(1):105-107. DOI: 10.1016/j.jns.2004.09.029. PMID 15607218